CONSERVATION CONCEPTS

This book provides a review of the multitude of conservation concepts, both from a scientific, philosophical, and social science perspective, asking how we want to shape our relationships with nature as humans, and providing guidance on which conservation approaches can help us to do this.

Nature conservation is a contested terrain and there is not only one idea about what constitutes conservation but many different ones, which sometimes are conflicting. Employing a conceptual and historical analysis, this book sorts and interprets the differing conservation concepts, with a special emphasis on narrative analysis as a means for describing human–nature relationships and for linking conservation science to practice and to society at large. Case studies illustrate the philosophical issues and help to analyse major controversies in conservation biology. While the main focus is on Western ideas of conservation, the book also touches upon non-Western, including indigenous, concepts. The approach taken in this book emphasises the often implicit strategic and societal dimensions of conservation concepts, including power relations. In finding a path through the multitude of concepts, the book showcases that it is necessary to maintain the plurality of approaches, in order to successfully address different situations and societal choices. Overall, this book highlights the very tension which conservation biology must withstand between science and society: between what is possible and what we want individually or as a society or even more what is desirable. Bringing some order into this multitude will support more efficient conservation and conservation biology.

This book will be of great interest to students and scholars studying nature conservation from a variety of disciplines, including biology, ecology, anthropology, sociology, geography, and philosophy. It will also be of use to professionals wanting to gain an understanding of the broad spectrum of conservation concepts and approaches and when to apply them.

Kurt Jax is Senior Scientist at the Helmholtz Centre for Environmental Research – UFZ in Leipzig, Germany, and Professor of Ecology at the TUM School of Life Sciences at the Technical University of Munich, Germany. He is the author of *Ecosystem Functioning* (2010).

Routledge Studies in Conservation and the Environment

This series includes a wide range of inter-disciplinary approaches to conservation and the environment, integrating perspectives from both social and natural sciences. Topics include, but are not limited to, development, environmental policy and politics, ecosystem change, natural resources (including land, water, oceans and forests), security, wildlife, protected areas, tourism, human-wildlife conflict, agriculture, economics, law and climate change.

Case Studies of Wildlife Ecology and Conservation in India
Edited by Orus Ilyas and Afifullah Khan

Creating Resilient Landscapes in an Era of Climate Change
Global Case Studies and Real-World Solutions
Edited by Amin Rastandeh and Meghann Jarchow

Species, Science and Society
The Role of Systematic Biology
Quentin Wheeler

Conservation Concepts
Rethinking Human–Nature Relationships
Kurt Jax

For more information about this series, please visit: www.routledge.com/Routledge-Studies-in-Conservation-and-the-Environment/book-series/RSICE

CONSERVATION CONCEPTS

Rethinking Human–Nature Relationships

Kurt Jax

Designed cover image: © Getty Images

First published 2024
by Routledge
4 Park Square, Milton Park, Abingdon, Oxon OX14 4RN

and by Routledge
605 Third Avenue, New York, NY 10158

Routledge is an imprint of the Taylor & Francis Group, an informa business

© 2024 Kurt Jax

The right of Kurt Jax to be identified as author of this work has been asserted in accordance with sections 77 and 78 of the Copyright, Designs and Patents Act 1988.

All rights reserved. No part of this book may be reprinted or reproduced or utilised in any form or by any electronic, mechanical, or other means, now known or hereafter invented, including photocopying and recording, or in any information storage or retrieval system, without permission in writing from the publishers.

Trademark notice: Product or corporate names may be trademarks or registered trademarks, and are used only for identification and explanation without intent to infringe.

British Library Cataloguing-in-Publication Data
A catalogue record for this book is available from the British Library

ISBN: 978-1-032-16922-4 (hbk)
ISBN: 978-1-032-16920-0 (pbk)
ISBN: 978-1-003-25100-2 (ebk)

DOI: 10.4324/9781003251002

Typeset in Times New Roman
by Newgen Publishing UK

CONTENTS

Acknowledgements   *vi*

1   Introduction   1

2   Situating conservation: definitions, origins, and context   7

3   Analysing conservation concepts   56

4   Western and non-Western ideas of nature and nature conservation   153

5   Moving forward: which conservation concepts for which purposes?   212

6   Conclusions and outlook   253

Index   *257*

ACKNOWLEDGEMENTS

This book has had a long period of development. It brings together many ideas and thoughts I have had on nature, conservation, and human–nature relationships throughout both my professional and personal life. Working on this book has in many ways been an experimental endeavour, of which I did not know where it would lead me. It is up to the reader to see if this experiment has been successful and if its results are useful. There are many people that have contributed to my thinking in one way or another. I cannot mention them all here, but I want to highlight those that have been instrumental in bringing this book together.

To start with, I would like to thank all the members of the working group "Theory and Science-Policy" (nickname: "Greyzone") at the Helmholtz Centre for Environmental Research (UFZ), where I discussed first ideas about structuring the multitude of conservation concepts some eight years ago. At the UFZ, the doors of these and other colleagues, from so many different disciplines – also beyond the natural sciences – have always been open, and I owe so much to the countless open and productive discussions with them. Special thanks go to Klaus Henle, former head of the Department for Conservation Biology at the UFZ, for providing excellent working conditions and leaving room for developing ideas beyond the mainstream of research. I would also like to thank my colleagues, at UFZ and beyond, with whom I worked in different projects and workshops. Special thanks to Barbara Muraca and Kai Chan who inspired and supported me to delve deeper into the concept of relational values.

Working in an interdisciplinary context has greatly widened my horizon and made my work truly enjoyable. In this context, I want to mention my late colleague Gabriele Brettschneider with whom I was working in the "Man and the Biosphere" programme in the 1980s; she opened the world of social sciences to me.

Thanks also go to my students at the Technical University of Munich, with their fresh and often unconventional questions, thereby helping my thinking about nature and nature conservation in different ways.

I would like to thank several people who read and commented earlier versions of chapters of this book, specifically Tina Heger, Thomas Kirchhoff, Ursula Schmedtje, Julian Rode, Ulrich Heink, Laÿna Droz, and Margarita (Maggie) Berg. Their critical comments on the manuscript helped to clarify my own lines of thought and avoid misunderstandings.

Finally, my very special thanks go to Ulla, for her support and her patience, having to deal with a husband who for a long time was deeply involved and sometimes totally immersed in the writing of this book.

1
INTRODUCTION

Nature conservation is contested terrain. It is so in many respects. It is contested between different parts of society with different interests on using and/or protecting nature. It is contested between different conservation practitioners about the right means to protect nature, but it also contested in terms about what nature conservation is (or should be) and thus with respect to its very foundations. There is, in fact, not only one idea about what constitutes conservation but many different ones, which sometimes are conflicting: in terms of the objects they aim at (e.g. ecosystems or species), in terms of underlying values, in terms of strategies to achieve it most effectively, and also in terms of the proper tools.

While concern for protecting nature and first practical efforts to do so were at first largely independent of science, conservation has been increasingly influenced by science, especially by biology and ecology. Since the 1980s even a specific research field, that of conservation biology, was established (Soulé 1985, Meine et al. 2006) with the development and use of specific scientific concepts, which at least since the "invention" of the biodiversity concept also made their way into various policy fields and even to the broad public. Especially prominent concepts are biodiversity, ecosystem services, but, even before that, concepts such as wilderness or, as supporting concepts, that of the "balance of nature" or "carrying capacity". Today one can see that these concepts are changing and succeeding each other at an increasing pace, in research as well as in policies. The impulse for writing this book came from just this observation.

During the last decades major conservation concepts (relevant to research and to policy) have followed each other more and more rapidly. Most recently, biodiversity was followed by ecosystem services and natural capital and now there is a new trend – at least in Europe – towards "nature-based solutions" and "green infrastructure", or – with the Intergovernmental Science-Policy Platform on

DOI: 10.4324/9781003251002-1

Biodiversity and Ecosystem Services (IPBES), a move from "ecosystem services" to "nature's contributions to people". My observation is that each new conservation concept prompts whole generations of scientists to jump on it and almost feverishly pursue research on the respective new concept, often abandoning old ones, either because the latter appear old-fashioned, have not proven "successful" (however that is assessed), or simply because funding sources have shifted to (sometimes only seemingly) new ideas. But what are we gaining, what are we loosing by such a succession of concepts? Do the new concepts include the old ones? Do they persist in parallel? Are some of them just old wine in new skins? And are they appropriate for saving nature, and more specifically *which* nature?

Looking at this is much more than an academic exercise. Instead, it leads beyond the mere theoretical analysis deeply into the very foundations of both conservation biology and conservation, or even broader to the question of what are (or could be) proper relations of humans with nature. While the starting point is conservation concepts, this book is thus not just about concepts but about values, aims, and strategies that underly and guide conservation and environmental management, how they interact with various societal discourses. It is, ultimately, also about our self-understanding as humans, which is related intimately to our ideas of nature and nature conservation.

Nature conservation is not monolithic. Although conservationists all strive to protect "nature", they mean highly diverse things when they say "nature" and have highly diverse motivations, goals, and values that underly nature conservation. The various conservation concepts, also those used in the scientific underpinning of conservation, are steeped by such values; they are mostly, even as it sometimes appears otherwise, not "neutral". It is these goals and values which are my focus, because only by knowing them can we also judge the success of the strategies based on particular conservation concepts.

Obviously, conservation is a societal activity, one among many others. It is supported by science but guided by societal choices of how we want to relate to nature. The first of the principles of the Convention on Biological Diversity's Ecosystem Approach therefore explicitly states "The objectives of management of land, water and living resources are a matter of societal choice" (UNEP/CBD 2000). Likewise, conservation biology from its outset was defined as "mission-driven discipline", embracing both facts and values (Soulé 1985).

There is thus a complex set of relations between conservation, science, and society. Research is guided by individual and societal interests (to preserve nature) and in itself aims at supporting conservation by scientific results, produced on the basis of scientific concepts and rigorous standards of "objective" investigation. These standards then are also meant to support conservation strategies and tools based on these scientific concepts and to produce results with strong credibility in society. It thus comes as no surprise that recent conservation concepts are more and more also "boundary objects", i.e. they are objects that facilitate communication across disciplines and even beyond academia by creating shared vocabulary although the understanding of the respective term could differ regarding its precise meaning (Star and Griesemer 1989).

Providing value-laden scientific concepts and boundary concepts has some advantage for communication between science, policy, and other parts of society, but it is also in danger of blurring differences and hiding the diverse and sometimes contradicting goals and values underlying the concepts. And the variety of goals and values is large, even among conservationists and among conservation biologists.

The variety and increasing succession of conservation concepts also causes confusion, not the least among policymakers. It is connected with confusion about what conservation actually is, and it can result in disunity, paralysis, weakening conservation against its opponents, unclear control of success of conservation outcomes, and sometimes unproductive activism.

One major reason for increasingly developing new concepts, working faster and faster, may be the perception how nature, how biodiversity is vanishing around us at an incredible pace and the feeling that we have no time left, that we, as humans concerned about the loss of important features of nature, need to act as fast as we can, doing whatever may help to save nature. Perhaps we are also too impatient to follow one line of argument and one major conceptual and strategic approach in the face of nature vanishing more and more. Another reason, in the academic field, is the high pressure to produce publications, produce results and be always "up to date" with a new concept that provides some taste of novelty, funding sources, and political goal fulfilment.

This book is an attempt to step back and to look at the multitude of conservation concepts, especially its underlying goals and values, but also the human–nature relationships implicit in them. The idea is to bring some order into this multitude, not as an end in itself but as support for more efficient conservation and conservation biology. The ultimate aim here is to search for clearer options when it comes to the question which is central not only to conservationists but to all of society: *How do we want to form (or reform) our relationship with nature as humans – and which conservation approaches can help us?*

This question has to be answered in the tension which conservation biology must stand in between science and society: between what is *possible* (as a core matter of science) and what we want individually or as a society – or even more: what is *desirable*. It also has to acknowledge that there is no simple "we" that embraces all humans but that different groups of people will have different ideas on what constitutes a good relationship with nature. As the focus here is on conservation concepts – not only, but mostly scientific ones – a major concern must also be on the role of science and of scientists in this complex relation.

Understanding the relations between conservation and society, as expressed in conservation concepts, is like facing a big intricate entangled knot, and there is no Alexander to cut it. We have to entangle it slowly and meticulously. We may start at different places: looking at the objects of conservation, at the practice, at the motivations, at the arguments posed, the values included, or at the main narratives included, the latter bringing together many of the former.

My approach to tackle these relations is an analytical one, trying to disentangle and then reorder the different meanings and different aspects of conservation concepts, namely their empirical bases, their conceptual ideas, the underlying values and strategies, but also the ideas of nature and those of desirable human–nature relationships implied. For this analysis I will also resort to the history of conservation; I will use this, however, mainly as a heuristic device, without claiming to be in any way exhaustive in terms of a complete history of the field. Nevertheless, this historic perspective is important. It helps to contextualise current concepts and the discourses related to them, and it lessens the danger to reinvent things time and again. My observation is that the history of ecology, conservation biology, and conservation is often too little known or is simplified too much to be really helpful. Furthermore, studying historical developments can create a bridge to the social context in which conservation is situated. Conservation is not only conducted by people but also affects people in various ways; therefore, the social context of conservation will also receive considerable attention in this book.

1.1 Structure of the book

In **Chapter 2** I will first define and delimit what I mean by conservation and conservation concepts and contextualise them, especially in their relations to science and to society. The chapter will also give a brief overview of the history of conservation and conservation biology.

Chapter 3 constitutes the analytical core of the book. Section 3.1 will introduce different ways by which one may distinguish (and classify) conservation concepts, e.g. via their target objects and the values embraced. As a tool to integrate different criteria, I will then focus on a narrative analysis of human–nature relationships as implicit in conservation concepts, using methods from literary studies, history, and other fields within the humanities (Section 3.2). To reconstruct such implicit narratives, a general structure of human–nature relationships in a conservation context will be developed. Building on this scheme, seven major narratives that can be found in conservation concepts will be described (Section 3.3). Finally, Section 3.4 will assemble a larger number of related conservation concepts into conceptual clusters (e.g. biodiversity, ecosystem services, cultural landscapes), at first sight partly linked to the narratives described before. For each of these clusters there is a brief discussion of its history, the commonalities as well as what distinguishes the concepts within a cluster, in terms of the objects, values, and narratives they refer to. At the end of that section, a comparison between the concepts is presented.

Chapter 4 explores some philosophical and anthropological background of conservation concepts and their differences. Starting with a discussion on "nature" and "natural" in Western thinking, it exposes the many different meanings of "nature" in Western culture (Section 4.1). Pointing at the many dichotomies on which these ideas of nature mostly build, I will ask how problematic they might be for conservation and how to handle them. As an extended case study, Section 4.2 will explore the question what it could mean to speak of nature as "socially constructed" and why this is very rarely something that has negative impact on conservation. While the book is focused mainly on Western ideas of nature conservation, it is necessary also to consider non-Western ideas of nature and conservation. This is important because Western ideas are often taken for granted and as universal to all humans. Section 4.3 provides some brief examples from non-Western cultures, demonstrating that and why this is an erroneous assumption. It also shows that – with all necessary caution – that we might learn from these other ideas for global conservation. The difficulties to "translate" or harmonise between Western and non-Western ideas of nature and its conservation are then taken up again by means of another case study (Section 4.4), analysing the debate about the creation of the Conceptual Framework for the IPBES. Finally Section 4.5 will briefly elaborate if or to what degree a narrative approach to conservation concepts, as highlighted in Chapter 3, may also be a tool to "translate" between Western and non-Western ideas of nature and nature conservation.

What does all this mean for selecting appropriate concepts for conservation? In **Chapter 5** I am trying to lay the ground for performing such decisions. I am at first pleading for restraint in proliferating too many new concepts, arguing that many of them are just revivals or reshaping of existing ideas. At the same time I argue for the use of multiple major conservation concepts, depending on the specific biophysical and social circumstances (Section 5.1). Extending the view, and as a necessary background for applying any conservation concept, the social and economic background of conservation has to be accounted for, as will be explained in some detail in Section 5.2. Building on and integrating the results from all previous chapters, Section 5.3 presents a set of criteria and requirements that should guide the selection of appropriate conservation concepts, based on the assumption that conservation aims at (re)constructing good human–nature relationships. The final section (Section 5.4) condenses these criteria into a brief proposal for a process allowing to clarify conservation goals and tackling conservation conflicts.

Chapter 6 provides a brief summary of the content and main conclusions of the book and presents some thoughts on the future of conservation.

References

Meine, C., Soule, M. and Noss, R.F. (2006). "A mission-driven discipline": the growth of conservation biology. *Conservation Biology*, 20, 631–651.

Soulé, M.E. (1985). What is conservation biology? *BioScience*, 35, 727–734.

Star, S.L. and Griesemer, J.R. (1989). Institutional ecology, translations and boundary objects: amateurs and professionals in Berkeley's Museum of-Vertebrate-Zoology, 1907–39. *Social Studies of Science*, 19, 387–420.

UNEP/CBD (2000). Ecosystem approach. In: *Decisions adopted by the Conference of the Parties to the Convention on Biological Diversity at its Fifth Meeting, Nairobi, 15–26 May 2000*. www.cbd.int/decision/cop/?id=7148

2
SITUATING CONSERVATION
Definitions, origins, and context

The purpose of this chapter is to introduce some basic definitions and distinctions as well as basic assumptions which underly this book. Even though some issues raised here will be fully discussed only in later chapters, introducing them here is necessary for understanding the full background of the approach and the context of the chapters that follow. For this purpose I will first ask for a suitable definition of conservation (Section 2.1) and then provide a short overview on the history of the field (Section 2.2), followed by some first thoughts on the meanings of "nature" and "natural" in conservation (Section 2.3). Delimiting conservation and conservation biology (as the research field dealing with it) against other fields of research and practice is the subject of Section 2.4. I then turn to the question of what is meant by conservation concepts (Section 2.5) and to the societal context in which these concepts and conservation in general are situated (Section 2.6). Section 2.7 opens up a discussion which will come up again at several places in the book, namely about the relation between science and conservation. The chapter closes with a case study about the debate on the "new conservation" (Section 2.8), which illustrates how some of the themes discussed before play out in the understanding and practice of conservation and conservation biology.

DOI: 10.4324/9781003251002-2

8 Situating conservation

2.1 What is conservation?

It is surprisingly difficult to find clear definitions of conservation, even in textbooks of conservation biology. David Ehrenfeld in his 1970 book *Biological Conservation*, which may be the earliest book with this title, wrote:

> Conservation is a broad topic, even when restricted to biological applications. There is no guarantee that people – including conservationists – are talking about the same thing when they refer to it.
>
> *(p. 2)*

So sometimes, conservation is seen as something that excludes humans and their needs, sometimes it is seen as including them, sometimes it is seen as referring only to nature as an end in itself, sometimes to nature serving human well-being. What Ehrenfeld stated in 1970 is still true today, as is evidenced, e.g., by the fierce debate on the "new conservation" (see Section 2.8).

In a recent article, Chris Sandbrook (2015) thus refutes the idea that there is one "right" and narrow definition of what conservation is. The broad definition which he uses comes close to my own understanding, and I will thus use it as my working definition. Sandbrook defines conservation as:

> actions that are intended to establish, improve or maintain good relations with nature.
>
> *(p. 565)*

This broad definition emphasises that conservation is about human action (including conscious restraint from acting) and it points at what I see as the overall goal of conservation, namely good human–nature relationships. It is, as Sandbrook points out, open to diverse objects of conservation and to either including people or not.

There are several areas where conservation grades into other fields, such as environmental protection, environmental management, or sustainable development, as well as other contested delimitations, e.g. between conservation and preservation (see later). Also there is sometimes a blurred boundary between conservation and conservation biology, the practice and the science underpinning it.

While I am dealing here mainly with concepts from conservation biology, a rather young discipline, many of these concepts have older "non-scientific" precursors, and it is necessary to view both older conservation ideas and more recent "scientific" conservation concepts. This is all the more important when the question is how much novelty is brought in by some of the latter and how the different ideas and concepts relate to each other.

Unless one is convinced that there is *per se* a duty to conserve "nature" in the state that it is or that it "originally" was (e.g. before humans arrived on the scene), conservation is clearly dependent on societal choices, based on the very idea(s) that we as humans have about good relations with nature. It is one – albeit important – action among many other human actions, such as economic, political, or everyday ones, with some of these other actions influencing conservation or being influenced by it, some of them not. The context of these various human activities has determined what and why people want to preserve as nature. This becomes clearer when taking a look at the history of conservation.

2.2 Conservation: a short historical overview

The brief journey into the history of conservation has two purposes. First, it serves to deliver the historical background of the conservation concepts that will be dealt with later, as a first glimpse on their origins. Second, such a venture also will clarify the scope of this study, both in time and space.

The regional emphasis will be on those regions where most current conservation concepts originated, i.e. the "Western" countries, including their colonies in the imperialist period (but see Chapter 4 for non-Western perspectives). Beyond North America, a special emphasis will be on the UK and continental Europe, not the least my home country Germany. These latter regions are often neglected in historical accounts of conservation and are interesting not the least with respect to protecting cultural landscapes as part of conservation. Surprisingly also, the early conservation movements were already much more international than studies focusing on just one or two countries would suggest. With respect to the temporal dimension of conservation, I will deal mostly with developments from the middle of the 19th century on, the time where conservation as an organised and planned activity began to gain traction.

The literature on the history of conservation is quite scattered. Some aspects have received special attention, e.g. the history of protected areas (Runte 1997, Sellars 1997, Sheail 2010, Gissibl et al. 2012a), specific ideas of conservation (especially wilderness: e.g. Nash 2014, Oelschlaeger 1991), or specific countries (e.g. UK: Sheail 1976, Evans 1997; USA: Norton 1991, Meine 2013; Germany: Knaut 1993, Ott et al. 1999, Schmoll 2004; France: Matagane 1998, Ford 2004; Russia: Weiner 1988; Canada: Loo 2006; Netherlands: van der Windt 2012). A somewhat broader approach is followed in Bill Adams' book *Against Extinction* (2004), but even here the focus is regionally restricted and mostly on large wildlife. An overarching global conservation history is not available yet, which is not surprising, given the breadth and diversity of the field. My short treatment here cannot and does not intend to substitute a more extended comparative study.

Origins in the 19th century and early 20th century

The roots of conservation, as well as its philosophical foundations and the motivations that led to its various forms, are complex and can be traced back many centuries (e.g. Haila 2012, Grove 1992, 1995). Thus, early protections of natural areas – especially forests – existed already since the Middle Ages and into early modern times, when nobility or the government preserved forested land as their hunting grounds or as a resource of timber for housing and war. Concerns on the consequences of deforestation – and even their influence on the (regional) climate – were already expressed in the 18th century (Grove 1992, Ford 2004). Many ideas preparing conservation go back to the age of the Enlightenment and the tradition of romanticism. The main developments of what we today call conservation, however, started in the 19th century, especially in a more organised way. Even though the conservation activities and organisations linked to it started at different decades in different countries, there were many commonalities, influences, and links between the developments in the different countries, and nature conservation soon became an international endeavour, with many connections and mutual inspirations between countries.

Two societal phenomena shaping the 19th century form the common background of the beginnings of Western conservation: the industrial revolution (including agricultural modernisation) and colonialism, coupled with imperialism. Both developments were also conducive to a strong increase in human population as well as (overall) improvements in living standards within the Western countries. In addition, the increasing mobility of people, together with a strongly growing interest in natural history, influenced the need for and the rise of conservation.

However, the idea that nature (or parts of it) had to be protected required a perception of an actual or at least imminent *loss* of something valuable or – in the case of animal protection – a new moral awareness regarding human behaviour towards sentient animals.

Important examples for observations of this loss were that of an over-collecting of plants and increasing incidences of cruelty against animals, both domestic and wild ones, the latter especially in the course of ("sport") hunting. Cruelty against animals was not only seen as creating harm to the animals themselves but also as harming human civilisation. As Sheal (1976, p. 9) expressed it for Britain:

> Such activities not only harmed wild life but human society itself, for the perpetrators were guilty of bestial, thoughtless or selfish acts. The protectionists believed that in saving wildlife, they were also helping to preserve the very fabric of society.

The same held true also for other countries (e.g. van der Windt 2012 for the Netherlands). The very idea that conservation was in many ways a duty for civilised societies was also visible in other areas, e.g. in the emerging idea of national parks (Gissibl et al. 2012a and see later).

Four major, partly interrelated strains can be described as the roots of conservation in the 19th century. These are (1) the protection of individual animals, (2) the protection of valued species, (3) the protection of landscape features (including organisms), and finally (4) the protection of whole landscapes. The motives for all these efforts were almost as diverse as they are today. As Ross (2015, p. 215) put it, "a potpourri of concerns" was underlying the various early conservation activities.

Protection of individual organisms

The oldest organised activity, which at least touches on nature conservation or can be seen as a precursor, was derived from concern about individual animals. The protection of animals (at first mostly domesticated ones) against cruelty in 1824 led to the creation of the first animal protection society, namely the Society for the Prevention of Cruelty to Animals in Great Britain, from 1840 on as the Royal Society for the Prevention of Cruelty to Animals (RSPCA) (Evans 1997, Sheail 1976). Similar organisations on a national level followed in other countries (e.g. France 1845, Switzerland 1861, USA 1866, Germany 1881; see Matagne 1998, Schmoll 2004, Davis 2015).

In a strict sense, animal protection is not regarded as nature conservation because its target is the well-being or welfare of *individual* organisms, but not populations, communities, ecosystems, or landscapes. I will therefore not follow this tradition in more detail, even though there are links, overlaps, and conflicts between animal protection and nature conservation (see Sections 2.4 and 3.1).

Protection of (particular groups of) species

At about the same time, i.e. in the late 19th century, organisations came into being whose aim was to protect specific species or species groups, e.g. birds, plants, and not the least "game" species, especially in the European colonies. The protection of bird and plant species was often seen as a kind of resource protection on various scales, largely without a focus on protecting specific areas, but also aesthetic, moral, and even functional ("ecological") concerns prevailed.

Birds were of especially important concern in many parts of the Western societies. In the UK, the Society for the Protection of Birds (SPB; since 1904 Royal Society, RSPB) was founded in 1889 (Sheail 1976). A few years later, also with the support of the RSPCA, the Sea Bird Protection Act was issued (1869) and then the Wild Birds Protection Act (1876). One reason for the creation of the RSPB was to combat the excessive use of bird feathers (not the least from tropical birds like Bird of Paradise but also from birds in Europe and the USA) for women's hats in the fashion of the time; women themselves were a driving force in the movement. According to van der Windt (2012) the use of feathers for millinery was a major reason for the foundation of the Dutch Society for the Protection of Birds in 1899. Similar organisations and laws for the protection of birds were established in several countries: in Germany, the *Deutscher Bund für Vogelschutz* (German Association for Bird Protection) was founded in 1899 by Lina Hähnle; a precursor in 1875 and a first national bird protection law were established already in 1888.

In the USA, the Audubon Society was founded on a national level in 1905, with precursors already from 1896 on; the Lacey Game and Wild Birds Preservation and Disposition Act was passed in 1900. The bird protection movement also soon became organised internationally, such as with the founding of the International Committee (later Council) for Bird Preservation (ICPB) in 1922 (today – since 1993 – BirdLife International).

Like in other areas of conservation, the reasons to protect birds were diverse. Hans von Berlepsch, an influential person in the bird protection movement, wrote in his popular book on bird protection:

> Bird protection is not just a hobby, a passion emerging from ethical and aesthetic motives – also not originating from the admiration of bird song, from the effort of beautifying and vivifying nature – but bird protection is primarily only an issue of national economy, an issue of eminent importance.
>
> Bird protection is a measure from which a material, a large monetary benefit accrues to us humans. Bird protection aims to protect and propagate the birds which are useful, yes directly necessary to humans.
> *(von Berlepsch 1899, p. 1f; translation KJ)*[1]

In consequence, Berlepsch distinguished between useful and noxious birds, and while providing practical recommendations how to preserve the "useful" birds in the major part of his little book, one chapter is also devoted to the "*Vernichtung der verschiedenen Feinde der zu schützenden Vögel*" ("The annihilation of the different enemies of the birds to be protected"). The latter group included cats, weasels, martens, but also sparrows, magpies, and a couple of other animals (von Berlepsch 1899, p. 73). This distinction between "good" and "bad" animals was not unique to German bird protection but a rather common one in early conservation (see Bildstein 2001 for an American example; Evans 1997, p. 30f for Britain). It even played an important role in the management of US national parks, in the form of predator control into the 1930s (Runte 1997, Sellars 1997); it does so even today, e.g. in connection with the question of how to deal with exotic and invasive species.

In contrast to this, there were also persons who objected to the utilitarian view on birds. For example, in Germany, the internationally well-known ornithologist Ernst Hartert argued for the protection of all birds (Schmoll 2005). Thus, the opinion developed that also "harmful" birds deserved protection. The bird protection movement, in fact, also embraced aesthetic, moral, emotional arguments as well as those related to the role of birds in the perceived "balance of nature".

As a consequence of this diversity of motives and the appeal of birds to so many different people, bird protection, at least in Germany, was a movement that brought together different social groups, from the political right to the left, from aristocrats to farmers and proletarians (Schmoll 2005) – in contrast to the mostly elite enterprises of many other facets of early conservation, such as game protection in the colonies and in the USA or early landscape protection in Europe (see later).

The protection of **large wildlife ("game species")** was connected to hunting – and overhunting – of large mammal species as well as of wild fowl. The experience of a decimation – and even extinction of species – was especially conspicuous in the second half of the 19th and the early 20th century, be it in the USA, with the near-extinction of the buffalo in the 1870s or the complete extinction of the passenger pigeon in 1899 (last animal in captivity to die in 1914), or in the UK, where several species like the brown bear and the wolf had already been exterminated long before. The observation was very prominent also in the colonies of the imperialist European nations in Africa and Asia (e.g. the extinction of the Quagga in the late 19th century). According to Jepson and Whittaker (2002, p. 133), around 1900, "the concept of human-induced extinction was established in the public mind".

Related to the protection of large wildlife, an important but still often neglected or even ignored root of conservation originated in the context of colonialism (see Box 2.1). The European colonial powers (especially the UK, France, Belgium, and – until 1918 – Germany) were indeed among the first to establish protected areas in their colonies, in the form of "game reserves". The background was the experience of decreasing abundances of large mammals in Africa and Asia, both from (partly illegal) commercial hunting, and "sport" hunting. These losses were then aggravated by the arrival of rinderpest in Africa, which heavily affected both domestic as well a wild mammals, such as buffalo and wildebeest (Adams 2004, p. 23f, Gissibl 2012, 104f). As a response to these developments, the colonial powers put in place hunting regulations (closed seasons, licences, etc.) and established game reserves. Several such reserves were created in the 1890s and the first decades of the 20th century, both in Africa (e.g. one of the earliest in 1896, a reserve around the Rufiji river in Tanzania) and in Asia.

Proposed by Hermann von Wissmann, former governor of German East Africa, a conference on game protection was held in London in 1900. Most of the African colonial powers joined in and a Convention for the Preservation of Animals, Birds and Fish in Africa was signed. Unfortunately, it was never implemented but it formed a blueprint for later conventions. The most important conference that built on the 1900 conference was held in 1933 and led to the adoption of the Convention on the Preservation of Flora and Fauna in their Natural State in the same year. McCormick (1989, p. 20) characterises this landmark convention:

> Superseding the long defunct 1900 Convention, it was designed to curb threats to African wildlife by creating protected areas, such as national parks and reserves. It brought preservationists, scientists and governments together in common cause, made its signatories aware of the problems of each other's African colonies, and established the precedent of nongovernmental organisations playing a technical advisory role in such initiatives. It even included appendices of endangered and rare animal species.

Hunters had a special role in all these developments. As Adams (2004, p. 30) states: "There is no doubt that the driving force for wildlife conservation at the start of the 20th century in both Africa and India were the European hunters". The influence of the hunters was not limited to these continents. Also in the USA, these elite groups had a major influence on the development of early conservation, with American and European hunters being in close cooperation (Jepson and Whittacker 2002). In the USA the major figure was Theodore Roosevelt, who, during his presidency, would establish a large number of national parks, national monuments, and wildlife reserves. Roosevelt, being an ardent hunter, created the famous Boone and Crockett Club, which was in close exchange with British hunter-conservationists such as Edward North Buxton, founder of the Society for the Preservation of Fauna in the Empire (SPWFE; founded 1903, now Fauna and Flora International).

The ideas behind the activities of this elite group also may give a clue to values that shaped not only game conservation in the late 19th and early 20th century but also the emergence of many types of protected areas. The obvious and immediate interest consisted in continued hunting opportunities, hunting being appreciated as a "manly" sports of European and American elites. Beyond that, however, protecting game and establishing game reserves in the tropics was seen as a matter of civilisation and at the same time one of national – and also global (see Ross 2015) – heritage. Edward North Buxton wrote:

> Game should be viewed as a precious inheritance of the empire, something to be most jealously safeguarded, like a unique picture – something which may easily be lost, but which cannot be replaced.
> *(Buxton 1902, p. 37)*

The comparison with a picture is telling as it alludes both to an aesthetic appeal and an elite view of nature.

As Grove (1990), Ross (2015), and others have emphasised, protecting the "wild" tropics was also a matter of idealised edenic ideas about nature, of "paradise" presumably still existing there, where it had already been lost in the highly industrialised European countries.

While the protection of game animals in the colonised countries was a strong force for an early internationalisation of conservation (Ross 2015, Gissibl 2012), it also carried with it highly problematic aspects (see Box 2.1). For instance, the view of "civilisation" was very elitist, gendered, and racist: it only considered white, Western civilisation and juxtaposed the noble European hunter with the hunting of the "cruel" indigenous people. As Adams notes:

> European hunters abroad took with them a long tradition of opposition to subsistence hunting. This was generally seen to be haphazard, inefficient, wasteful and cruel.
> *(Adams 2004, p. 31)*

BOX 2.1 COLONIALISM AND NATURE CONSERVATION

The colonial and imperialistic legacy of conservation is often ignored or forgotten. However, there are good reasons to deal with this subject, not only from a historical perspective but also as relevant into the present.

On the positive side, colonialism led to early ideas and practices of wildlife protection and even to a first internationalisation of this direction of conservation (Gissibl 2012). As described earlier, motivations for this kind of imperial conservation reached from the safeguarding of hunting grounds for the Western elites via the prevention of indigenous hunting practices that were perceived as cruel to the early notion of a – first national, then global – heritage of an "untouched", "wild", even "edenic" nature that was already gone in the home countries of the imperial powers. But his came at a high price for the local human populations:

> The global conservationist view that motivated the preservation of wildlife in East Africa was, therefore, not only informed by idealist motivations. It was detached, temporalized, gendered and racialized too.
> (Gissibl 2012, p. 108)

Based on a Eurocentric idea of "civilisation", indigenous people, but also white settlers, with their respective practices of interactions with wildlife, were perceived as inferior to the white (male) elites. In consequence, these people were either killed, displaced, or their original ways of life were strongly restricted, with severe consequences for local societies (Grove 1990). In the form of what some people have called "internal colonialism", this happened also beyond the European colonies, e.g. in North America, where the establishment of some of the early National Parks led to the displacement of its original native inhabitants (Spence 1999).

But the old imperialistic thinking is not just a matter of the past. As Agrawal (1997) showed, old ideas of progress, civilisation, and the superiority of Western knowledge and thinking prevail in conservation. Conservation (implicitly) based on such assumptions may not always be executed by brute force today, in terms of evicting people from their dwelling places (but see Brockington and Igoe 2006). More often it is expressed in imposing Western ideas, with various social and economic consequences. Protected area concepts, especially those restricting their traditional activities, are a special subject of dissent between conservationists and indigenous people (e.g. Lee 2016 for Australia). The imperialistic nature of some conservation approaches may not be conscious to and intended by its supporters. Awareness on this issue and its prevailing high importance also in today's conservation must be a matter of concern for conservationists (see Section 5.2).

16 Situating conservation

This referred both to the indigenous peoples and their traditional hunting methods as well as to the European settlers (and, of course, poachers).

The aim of protecting **plants and plant communities** did not result in specific conservation organisations on a national or international level. Even though plants were sometimes included in conservation programmes (e.g. in the name and programme of the Selborne Society for the Protection of Birds, Plants and Pleasant Places, founded in 1885 and considered a predecessor of the RSPB), plants (e.g. old trees) mostly were subsumed as parts or special features of landscapes.

Protecting remarkable natural features

In the densely settled landscapes of Europe it was not – as in the USA – the vast wild and scenic landscapes that were the major conservation target. Instead, often much smaller, remarkable, and rare features of nature were protected. As with large game in Africa and elsewhere, this aspect of conservation was partly motivated by protecting national or global heritage. This is well-manifested in the idea of "natural monuments".[2] In Germany, the start of conservation is considered to be the designation of the Drachenfels mountain (at the banks of the river Rhine, south of Bonn) as a *Naturdenkmal* (natural monument) in 1836. The Drachenfels, a basaltic hill with the ruin of a castle on top, which was in danger of being destroyed by a quarry at its western flank (Figure 2.1), was both of natural and cultural significance, connected with legends, Rhine romanticism, and national

FIGURE 2.1 The Drachenfels hill at the banks of the river Rhine. Photo: Kurt Jax.

symbolism. Later, natural monuments were mostly remarkable, peculiar or rare features of nature alone, like old trees or remarkable rock formations, "witnesses" of the past of the specific country and region. Also, these landscape elements were more and more endangered by the rapid industrialisation of the 19th and early 20th century. The desire to protect them was thus often also a critique of these developments.

The concept of the natural monument was brought into the mainstream of the emerging conservation movement most significantly by Hugo Conwentz, a German botanist. In 1904, Conwentz published a highly influential memorandum with the title "*Die Gefährdung der Naturdenkmäler und Vorschläge zu ihrer Erhaltung*" ("The endangerment of the natural monuments and proposals for their preservation") and in 1906 he became the first head of the "Prussian State Agency for the Care of Natural Monuments" (a predecessor of the current Federal Agency for Nature Conservation, BfN). In a later publication on this topic he wrote:

> Under the term *Natural Monument*, characteristic formations of nature are understood, particularly those which are still in their primitive location and have remained completely, or almost completely, untouched by civilisation. To these belong not only individual forms and species, but also plant and animal associations, geological and scenic rarities.
> *(Conwentz 1914, p. 110)*

The concept was applied to various features, both abiotic and biotic, also to rare species and even larger parts of the landscape. There was thus an overlap with the protection of whole landscapes (see later). Conwentz also acknowledged that in densely populated countries such objects which are completely "untouched by civilisation" are often not found and that, therefore, the term "natural monument" was also used in a broader sense.

The idea of national monuments obviously was a timely one as societies for the protection of natural monuments formed in several European countries, such as France (1901), the Netherlands (1905), Switzerland (1909), or the UK (1909) (see Jepson and Whittaker 2002, Ford 2004, van der Windt 2012), partly under influence of Conwentz and using his guidelines for the assessment and protection of such "monuments" (Ross 2015).

In the USA, the "Antiquities Act" was passed in 1906, after which President Roosevelt designated several "national monuments", among them large areas such as the Grand Canyon, which in part were later transformed into national parks. According to Kupper (2014) no direct influences from the European conservation movement could be established from the literature, so it may be more of a coincidence, due to the general zeitgeist.

The motivations for protecting natural monuments were again diverse. Overall, many protagonists saw it as a "national and patriotic duty" (Conwentz 1914, p. 119; see also Ford 2004 for France), to preserve these remarkable and rare features of nature also for the next generations (heritage argument). Overall, however, there were a number of other reasons why particular natural objects were to be preserved. These were partly scientific reasons, namely "to determine what changes have occurred in a district which had been [as a conservation measure] entirely abandoned to itself"; in another case natural monuments were meant to serve as a "sanctuary for birds" (i.e. the habitat is protected for and with the birds) and even "to preserve model forest tracts for landscape painters" (all quotes Conwentz 1914, p. 115).

The concept of natural monuments still exists today and is included by IUCN as category III in its protected area scheme (see Dudley 2013). The protection of rare and remarkable features is still high on the conservation agenda. Likewise, the protection of natural (and/or cultural) objects as global human heritage is formalised in UNESCO's World Heritage Convention (see Section 3.4).

Even though quite popular in the late 19th and early 20th century, the concept of only preserving more or less isolated natural features of the landscape was also criticised by many early conservationists. Criticism focused on the small size of most natural monuments, arguing that whole landscapes should be protected – with or without its human inhabitants and their cultural influences.

Protecting whole landscapes

Different types of larger areas were considered worth preserving, and again for various reasons. These protected areas included areas safeguarded for their natural resources (game, trees, etc.), or habitats for valued species (e.g. birds), as well as larger natural monuments, areas protected as/for wilderness, national parks (of different kinds and purposes), and also cultural landscapes, in which nature and culture were protected as "harmonious" units. The motivations to protect these different areas were not as clear-cut as it seems from some of the titles they were given or how they were depicted in the literature. Quite frequently multiple reasons existed for protecting a particular place, the reasoning often changing over time. Already the description given by Conwentz (see previous section) for the purposes of protecting larger natural monuments shows this variety.

The earliest protected areas go back to **game reserves and reserves to protect timber and other forest resources**. In a way, this kind of protection was based on what would be called scientific or ecological arguments today, the protection – and even restoration – of forests in order to prevent erosion and negative effects on the (local or regional) climate. As Grove (1992) has described, this happened, e.g., already in the early 18th century, when the French, guided by scientists, tried to reforest parts of the island of Mauritius. The reasons were economic, but also aesthetic and moral feelings, and – importantly – early insights into the relations between deforestation, soil erosion, and local climate change. Other colonial powers, e.g. the British, established forest reserves based on "ecological" reasons in their colonies already in the middle of the 18th century. In the USA, George Perkins Marsh in 1864 was arguing for the protection and even restoration of forests. Forestry science was among the earliest to develop systematic management rules. Also, it was forestry in which the concept of sustainability was developed as an early version for prudent resource use. The term sustainability (German: *Nachhaltigkeit*) – or initially the adjective "sustainable" (*nachhaltig*) was first used in 1817 by the Saxonian forester Hans Carl von Carlowitz in his book *Silvicultura Oeconomica* to describe the need for a continued usability of forests. In the 19th and early 20th century, German forestry science was famous and influenced that of many other countries. Thus, also Gifford Pinchot and Aldo Leopold, two icons of the American conservation movement, both visited Germany to study forestry science. The dominance of a resource-oriented, mainly utilitarian forestry led to a major division especially in the American conservation community, with a juxtaposition of conservation in a narrow sense (as resource use-oriented) and preservation (as oriented towards protecting nature without direct use by humans). I here subsume both approaches under the umbrella term "conservation" (see also Section 2.4). However, not all forests were set aside either for sustained resource use or on behalf of protecting "wild" nature for its own sake. Thus, the first protected forests in France (such as the forest of Fontainebleau near Paris, protected 1861) were conserved in fact also for symbolic and aesthetic reasons, as a national heritage, a national monument. Different groups of society emphasised one or the other of these reasons, resulting also in conflicts (Ford 2004).

Obviously, the focus on the type of landscapes that were to be protected varied with the specific country and with the respective type of landscape that was seen as being endangered. Besides near-natural areas, such as forests, there were also cultural landscapes in Europe (and some remnants of "pristine" landscapes there), while in North America, and also in the European colonies, the focus was on "wild" landscapes and the perceived "pristine" wilderness.

To a large degree, both **wilderness areas and national parks** as types of protected areas were an American invention. The two types of areas should not be confounded, however, because, although the idea of wilderness and the creation of the first national parks had close links, there were and are also significant differences.

The first national park was founded in the USA in 1872, namely Yellowstone National Park (Figures 2.2 and 2.3). The concept of the national park was, from the onset, connected with multiple ideas (see later). There are many excellent works on these ideas (e.g. Runte 1997, Sellars 1997, Sheail 2010, Gissibl et al. 2012a), and I may not be able to do justice to them in my short sketch.

The founding act of Yellowstone[3] describes the purpose of the park as being "set apart as a public park or pleasuring-ground for the benefit and enjoyment of the people". Yellowstone was created because of its scenic beauty, its "natural wonders" (not the least the geysers and other thermal features), and its abundant wildlife, but it was not created as a total reserve. The emphasis in the years to come was in fact strongly on tourism, recreation, and a general experience of wildness

20 Situating conservation

FIGURE 2.2 Yellowstone National Park. The inscription on the Roosevelt Arch, forming the northwestern gateway into the park, is a quote from the 1872 founding law of the park. Photo: Kurt Jax.

FIGURE 2.3 Yellowstone National Park is inter alia known for its abundant wildlife, including large mammals such as wapitis (*Cervus elaphus*). Photo: Kurt Jax.

for "tired, nerve-shaken, over-civilized people" (Muir 1901/2019, p. 1), much less than protecting nature for its own sake. Early American national parks were mainly protected for the reasons mentioned earlier, connected to a "cultural nationalism" (Runte 1997), similar to the conservation ideas developed at the time in European countries, although the latter were mostly applied to natural monuments and cultural landscapes. When I first visited Yosemite National Park in 1989, I was struck by the amazing movie shown in the visitor centre and not the least by its motto, meant to motivate people to protect the park: "Pride in America". So the national impetus that had been so important in the early days of conservation was still an argument into the late 20th (and 21st) century. As Karen Jones (2012, p. 35) expressed it: "Nature was nationalized and nation naturalized". Beyond that, a longing for a lost paradise of "untouched" nature found its expression in the establishment of the early national parks, with an idea of "pure" nature as free of human intervention – sometimes to the detriment of native American people (Spence 1999).

For still others, most prominently John Muir (1838–1914) who was often called "the father of national parks", going into the wild and arguing for national parks was about much more than recreation. For Muir it was a deeply spiritual, religious experience (see Worster 2008). This view comes closest to the later idea of protecting nature for its own (or God's) sake. Muir was the driving force behind the designation of Yosemite as a state park by the state of California in 1864 (becoming a national park in 1890), and he also was the founder of the Sierra Club (1892) as the oldest "grass root" conservation organisation of its kind in the USA. It should be mentioned, however, that Muir also advertised the advantages of wilderness for tourism and (mental) recreation as a strategic argument for promoting the establishment of national parks.

The concept of the national parks was disseminated across the world, first of all in those countries with settler communities of European origin, such as Canada, Australia, and New Zealand (Jones 2012). In Europe, the first national parks were created in Sweden (1909) and Switzerland (1914), in Africa the first were in Algeria (1923, see Davis 2012) and then Belgian Kongo (1925). But even though the first national parks were created in the USA, there was at first not a full-fledged idea of what a national park should be, nor was there a network of parks and a general management scheme (Jones 2012, Tyrrell 2012) or an official state agency to manage them. The first systematic network of national parks and a corresponding agency was established in Sweden from 1909 on, while the US National Park Service was founded only in 1916, five years later than the Canadian Parks Branch (1911).

In contrast to the focus on recreation and education (in continuing tension with the other goal of preserving of what was perceived as "pristine nature"), as given in the early US national parks, the – first and up to now only – Swiss national park was one established with a completely different purpose, mainly as an almost total reserve, serving as an area for scientific research (Kupper 2009).

The different ideas of what are the main criteria and aims for national parks, or what is called "national park" in different countries, persist to this day. This was even not solved by the effort of the International Union for the Conservation of Nature (IUCN) to classify protected areas, beginning in the 1960s. While the original idea of parks like Yellowstone was encoded in IUCN's category II, "areas with the national denomination of a 'national park' became distributed over all IUCN categories, whereas areas that met the criteria were included in Category II although they do not bear the name 'national park'" (Gissibl et al. 2012b, p. 13).

Regarding the relation of national parks to **wilderness**, there were also major differences (see Section 3.4 for more details on the wilderness concept and its history). In spite of Muir's efforts to promote wilderness in the USA, Runte (1997, p. 60) states that "not until the 1930s would wilderness preservation be recognized as a primary justification for establishing national parks, at least in the eyes of congress"; the "Wilderness Society" was founded only in 1935. Strict wilderness areas, where humans and especially human infrastructure are almost completely excluded, were in fact contradicting the setup of most early national parks and even today form a different category in IUCN's scheme of protected areas. A system of areas devoted exclusively to wilderness preservation ("wilderness areas", IUCN category Ib) was, e.g., formally established in the USA only by the Wilderness Act of 1964.

The contrast made up between preservation (in national parks) and conservation (in national forests) led to major controversies in the USA (see Section 2.4) but much less so in the European countries. Many historians today state that "wilderness" was an invention of Western civilisation much less than a reality perceived by local (indigenous) peoples living in or near presumed wilderness areas, even in places like Amazonia (e.g. Mann 2005; see also Sections 3.4 and 4.3). Most landscapes were early on inhabited and – to different degrees – in fact cultural landscapes.

Especially within the densely settled European countries the idea of **cultural landscapes**, moulded by a long history of harmonious interaction between humans and nature, was explicitly one of the major starting points of conservation. Also here, a feeling of loss was a major driver for people to postulate their conservation. Far from mere resource protection, aesthetic, symbolic, and experiential dimensions of various kinds shaped the conservation of cultural landscapes, among them even explicit rejection of a purely utilitarian, use-oriented perspective. Urbanisation and industrialisation, especially in the agricultural sector, as well as – in some countries – increasing tourism were threatening the traditional cultural landscape in the eyes of many early conservationists, and with it also threatened the national character of these areas. This national character for them was not expressed by the natural setting alone but both by (traditional) nature *and* culture, which was seen as forming a harmonious unit. This direction of conservation was followed in an organised manner in several European countries such as Germany, the UK, France, Switzerland, Italy, Denmark, Sweden, or Belgium (Schmoll 2004, Schlimm 2020). Its major focus was on the preservation of *national* heritage, but there was also an emphasis on *local* and *regional* heritage as shaping people's identity. Sometimes, e.g. in France, even the perspective of the common heritage of humankind was evoked (Schlimm 2020). While there were two international conferences on the protection of landscapes before WWI (1909 in Paris, 1912 in Stuttgart), the specific emphases of landscape protection within the different countries, and the major problems addressed, differed. In the UK, e.g., a major theme had long since been the question of access to the countryside, leading to the creation of the Commons Preservations Society already in 1865 and the National Trust for Places of Historic Interest or Natural Beauty in 1895 (in spite of its name, being a private organisation). A major goal of both organisations was to preserve Britain's cultural and natural heritage and guarantee continued access to it (Sheail 1976, Evans 1997). Among the driving forces behind these efforts were ramblers, sportsmen, and tourists. Because the acquisition of land by the National Trust was considered as somewhat haphazard from the perspective of naturalists and scientists, the Society for the Promotion of Nature Reserves (today Royal Society of Wildlife Trusts) was created in 1912 under the initiative of Nathaniel Charles Rothschild. The Society produced a systematic list of areas "which retain primitive conditions and contain rare and local species liable to extinction" (cited in Evans 1997, p. 46), which should be acquired and then protected as reserves (after having been handed over to the National Trust).

In France, building on earlier efforts, the Society for the Protection of Landscapes (*Société pour la Protection des Paysages de France*, today *Société pour la protection des paysages et de l'esthétique de la France*) was founded in 1901 (Matagne 1998). Here, as in other countries, aesthetic and symbolic (not the least nationalist) reasons predominated this movement, and "[f]orests and plant life were not valued in and of themselves, but rather as an expression of the past, of history" (Ford 2004, p. 182f).

In Germany, the protection of cultural landscapes is even considered to form a major starting point of nature conservation ("*Naturschutz*"). Its main initiator was Ernst Rudorff, a musician from Berlin. For Rudorff, who is said to have coined the term *Naturschutz* (presumably 1888; Knaut 1990), it was part of the broader idea of "*Heimatschutz*". *Heimatschutz* means the protection of the home country or home land, the "*Heimat*". *Heimat* is a complex concept, intimately related to people's identity and sense of belonging and it embraces (traditional) culture as well as nature, in their "organically grown" relationship, and a "holistic" understanding of a unity of humans and nature. Rudorff was the main proponent of *Heimatschutz* in Germany, publishing several programmatic texts on the idea (especially Rudorff 1904; first edition already published in 1901) and was the driving force behind the creation of the German Association for Homeland Protection (*Deutscher Bund Heimatschutz*) in 1904. As part of this broader approach, *Naturschutz* aimed at the traditional, pre-industrial landscape, also with its traditional land use practices. The protection of parts of primordial ("*ursprünglicher*") nature or even wilderness was not excluded but never formed a major emphasis of early *Naturschutz*. Instead, the focus was on cultural landscapes (*Kulturlandschaften*), even though Rudorff to my knowledge never used the latter term (see Section 3.4 for more on this concept). The protection of nature here was also one of aesthetics but also meant to have a civilising or even positive moral ("purifying and uplifting"; Rudorff 1880/1990, p. 122) effect to those enjoying it in a proper way. The notion of *Heimat* and with it *Naturschutz* was mostly set in a politically and socially conservative understanding of a "natural order".

Schmoll (2004) states:

> "Heimat" suggested the human environment as the natural order and declared rural life as natural life.
>
> *(Schmoll 2004, p. 388, translation KJ)*[4]

Heimatschutz was promoted at first by an (urban) elite, part of which even wanted the peasants to continue their traditional ways of life and not open themselves to modernity. Rudorff, e.g., even rejected the use of machines in traditional farming, arguing that it was not needed and that the health of farmers was served better if they used

> the emptiness of many an hour of the long winter to flail their grain, which will keep the energy of their muscles fresh, instead of them shambling out of boredom to the next train station to search for urban amusements.[5]
>
> *(Rudorff 1904, p. 79f, translation KJ)*

This conservative and often backward-looking direction of *Naturschutz* strongly emphasised national as well as regional and local identity (see Lekan 2004 for a detailed discussion), trying to bring together people and the still young nation in a common German heritage.

Even though not necessarily related to far-right nationalist ideas links to such ideas brought the conservation movement in Germany in discredit in the second half of the 20th century. The main reason for this is that at least parts of the movement (e.g. Paul Schulze-Naumburg, the successor of Ernst Rudorff in the *Bund Heimatschutz*) actively sided with Nazis and their blood and soil ideology. In the first decades of post-war Germany, this led to a very reluctant attitude towards nature conservation in general and the notion of *Heimat* in particular, especially on the political and intellectual left (see Radkau and Uekötter 2003).

Although the protection of cultural landscapes has been important in bringing conservation into the public minds in late-19th- and early-20th-century Europe, and although several organisations mentioned earlier still exist, the importance of these *organisations* in the conservation community is relatively minor today. One important reason will be discussed in the following section, namely the increasing influence of science during the 20th century, which was largely absent to the landscape protection movement. Landscape protection was much more focused on aesthetics and the various human relations with nature, material as well as symbolic ones, than on science. Nevertheless, the notion of cultural landscapes, without the conservative political burden it gained during the nationalist era, continues to be an important approach to conservation in settled landscapes, in Europe and also far beyond Europe – perhaps more than ever (see Droste et al. 1995 and Section 3.4).

Early conservation: some major commonalities

The conservation movement was (and is until this day) highly fragmented. As elaborated earlier, the objects of conservation as well as its motivations were highly diverse. There are, however, some major commonalities throughout these diverse ideas and types of conservation, which can help to understand also current arguments on conservation.

First of all, common to almost all approaches is the perception of already occurring or at least imminent losses, to which conservation measures should react. Among the justifications why something should be protected, a few stand out.

There is, of course, the protection of resources and the avoidance of catastrophes, e.g., through deforestation, problems described succinctly already by Marsh (1864) but also by early scientists. But even more prominent were the recreational and symbolic qualities of nature. The symbolic relevance of nature referred both to national, regional, and local identity as well as pride in one's own country and its natural (and cultural) heritage. In terms of recreation, the experience of aesthetic and sublime landscapes and objects within the landscapes was seen not only as relaxing the stressed urban traveller but also as uplifting the soul and as improving the morals of people, i.e. also as a social benefit. Directly linked to that was the idea of conservation as sign of or even a duty of civilised people, be it to avoid cruelty to animals or for protecting national or global heritage. Also, ideas of edenic, pristine, "untouched" nature as true nature – as a contrast to the spoiled industrial and urban landscapes – formed the background motivation to preserve nature as an act of a civilised nation. As Ross pointed out:

> Paradoxically, the threat posed by the "advance of civilization" meant that nature protection became part of the civilizing mission.
>
> *(Ross 2015, p. 220)*

These ideas did not follow each other in a temporal sequence but existed at the same time and were sometimes even applied to the same objects and places. John Muir and Ernst Rudorff (see Figures 2.4 and 2.5), e.g., were almost exact contemporaries, publishing their main programmatic texts almost at the same time. Muir (1901) was arguing for protecting wilderness and Rudorff (1901) for protecting cultural landscapes.

What was almost absent in these early mainstream conservation efforts was both the idea of protecting nature for its own sake as well as explicitly religious

26 Situating conservation

FIGURE 2.4 (left)   John Muir. Photo: Carleton Watkins (ca.1875). Public domain, via Wikimedia Commons.

FIGURE 2.5 (right)   Ernst Rudorff (before 1892). Photo: August Weger, Public domain, via Wikimedia Commons.

motivations for conservation (however, with few notable exceptions such as John Muir). Also, scientific underpinnings of conservation were rather rare at first, even though in part scientists engaged in setting aside some "pristine" areas as reference sites for research (e.g. the Swiss National Park). This would, however, change during the 20th century (see later).

There were few real "conservation concepts" in the sense of programmatic ideas for conservation, formulated at the time, but most *ideas* which we deal with today were already around, mostly without them being formally seen as "conservation concepts". One can, e.g., see prudential arguments referring to the use of natural resources; there was the idea of wilderness and even that of restoration. Beyond that, we find the notions of cultural landscapes and that of global heritage. Also, different types of protected areas, a hallmark of 20th-century conservation (Adams 2004, p. 4), were already proposed and implemented early on. Until the 1940s most ideas and arguments for nature conservation were in place, even if not expressed in today's terms.

Conservation after 1945

I have devoted some space to the early history of nature conservation, as many ideas evolved during this time and are still highly influential today, even though sometimes in a hidden manner (e.g. the colonial background of conservation; see Section 5.2). I will, however, not follow the further development of conservation organisations and laws after WWII in such detail but only sketch the most important trends and new ideas on conservation as they developed in the following decades and until today.

Overall, this section describes and discusses two major trends in the development of conservation: one is the increasing internationalisation of conservation, the second is the growing influence, today almost dominance, of science in conservation, with a formalisation and reconfiguration of conservation ideas in the 20th and early 21st century.

The rise of international conservation

As shown earlier, conservation already in its beginnings did not consist of isolated national endeavours. In a sense they were multinational, as the approaches in the different countries influenced each other, with ideas and strategies being exchanged in international (mostly European) conferences. But it was not international in the sense that multilateral agreements obliged member states towards the same goals (as, e.g., today the Convention on Biological Diversity). One exception was the Convention for the Preservation of Animals, Birds and Fish in Africa (London 1900) that, however, was never ratified, or only in a modified way at the follow-up conference in 1933 (see earlier). Another early effort for internationalising conservation was the First International Conference for the Protection of Nature, in Berne, Switzerland in 1913. At this conference, its initiator, Swiss zoologist Paul Sarasin, suggested to establish an international organisation for nature protection. A foundation act for such an organisation was even signed by 17 countries at the Bern conference, but the outbreak of WWI eventually stopped the project (McCormick 1989, Ross 2015). In 1928, an International Office for the Protection of Nature was created, which, after WWII (1948), in effect was merged with the new International Union for the Protection of Nature (IUPN; since 1956 IUCN) (Talbot 1983).

Large international institutions for nature conservation thus mostly came into being after the WWII (see Curry-Lindahl 1978, Adams 2004 for details): IUCN, UNESCO (founded in 1945; especially its programm Man and the Biosphere, from 1970 on), UNEP (1972), much later the Convention on Biodiversity (CBD) and the Intergovernmental Science-Policy Platform on Biodiversity and Ecosystem Services (IPBES) as well as some smaller but nevertheless important agreements such as the Ramsar Convention on Wetlands (1971) and the Bonn Convention on the Conservation of Migratory Species (1979). Also large NGOs formed at that time (Nature Conservancy 1950, WWF 1961). Some of these also developed from institutions which at first worked on a national basis but later extended their scope (e.g. the Nature Conservancy). In addition, new concepts that did not come from within conservation but touched upon it, such as sustainable development, were introduced (see Section 2.4).

It should be noted here that especially the early 21st century saw an increasing trend towards cooperations between international conservation organisations and large for-profit companies, a process which has been heavily criticised by parts of the conservation community and especially by political ecologists (e.g. Büscher and Fletcher 2020; see also Sections 2.8 and 5.2).

The increasing influence of science on conservation

Science strongly increased in importance for conservation after WWII. Scientific support had been negligible for much of early conservation (with the notable exception of forestry and forest protection), either because it was ignored or not available. What van der Windt states for the Netherlands around 1900 might be seen as typical for the science of the time:

> [A]round the turn of the century, many scientists considered nature conservation to be neither an important nor a necessary activity for scientists. For most biological organisations, science and nature conservation constituted two different discourses.
>
> *(van der Windt 2012, p. 218)*

In some countries and some selected themes of conservation, science played a role, such as in the German protection of natural monuments. The focus here, however, was more on botany and zoology of single species, and to some degree plant geography, and much less on ecology in the modern sense. An exception might be the establishment of the Swiss National Park in 1911 (see earlier), in which plant sociologists, aiming at whole plant communities instead of single species, played an important role.

Ecology, as the scientific discipline that was obviously especially qualified to support conservation, became a "self-conscious science" (Allee et al. 1949) only during the 1880s (McIntosh 1985, Kingsland 1985, Jax 2011) and was for the most part not devoted to problems of conservation yet. Ecologists' highly ambivalent attitude to research on and even more on the active promotion of conservation is shown, e.g., by the fate of the Committee on the Preservation of Natural Conditions of the Ecological Society of America (ESA). Animal ecologist Victor Shelford was one of those ecologists highly engaged in conservation. In 1917 he created the Committee within the then still young Society (established in 1915). The goals of the committee were diverse, but an important role consisted in securing natural, "pristine" reference areas for ecological research (Kinchy 2006). Not everybody welcomed these activities and eventually, in 1945, the society decided that direct action on preservation by the Society would be prohibited. This led, in 1946, to the formation of a new, independent organisation, the Ecologists' Union, which in 1950 became today's Nature Conservancy (Kinchy 2006). In the management of US National Parks as well, science had a varying and initially negligible influence, with some ups (first in the 1920s) and downs (e.g. in the New Deal era), and at first mainly with respect to wildlife management (Sellars 1997, Kupper 2009). It was Aldo Leopold (Figure 2.6), who, since the early 1930s, explicitly introduced contemporary ecological concepts into conservation. Departing from his earlier single-resource-oriented focus, which he had been trained on as a forester and game manager, Leopold in particular turned towards the community concept, inspired by

FIGURE 2.6 Aldo Leopold (1944). Copyright: Zhagen2024.
(https://commons.wikimedia.org/wiki/File:Aldo_Leopold_Studio_portrait,_1944.jpg), https://creative
commons.org/licenses/by-sa/4.0/legalcode.

British animal ecologist Charles Elton, considering it as a major concept relevant for conservation (Leopold 1939, 1949).

In Britain, early conservation was not much a subject of ecologists either, even though the British Ecological Society, founded in 1913, was the oldest ecological society in the world. While ecologists offered some support to pest management, forestry, and pasture management (partly in vain), there was at first no research aiming at conservation which went beyond securing natural resources. Writing about the year 1941, Bocking (1993, p. 97) states: "Ecologists did not at this point play a major role in the nature reserves movement, possibly preferring to leave the initiative to amateur naturalists and recreational interests". Only in the 1940s did this interest increase, partly on behalf of pre-existing aesthetic and heritage motives of the ecologists, but even more with the aim of securing protected areas for ecological research and securing government funding for this research; the British government only in the early 1940s had begun to show a greater interest and active role in the nature reserve movement (Bocking 1993). Thus, science was not only seen as supporting conservation but also vice versa: conservation helped to support ecologists' "professional agenda" (Bocking 2020). A driving force here was eminent ecologist Arthur Tansley, well known for coining the ecosystem concept. In contrast to their American counterparts, the focus of British ecologists was explicitly not on pristine but on managed, cultural landscapes, as it was evident that most British ecosystems had been influenced by human management and also needed further – at best science-based – management.

The 1960s are generally seen as the beginning of the broader environmental movement with one major landmark being the publication of Rachel Carson's book *Silent Spring* (1962). In this decade, ecology was pushed for the first time into the limelight of society whereas it previously had been a kind of obscure and exotic field of biology, beyond the dominant lab science. From now on, high expectations both from governments and, even more, the general public were addressed to ecologists. These came on the one hand from a technical side of guiding the management of natural resources and the conservation of species and ecosystems and on the other from the upcoming green movement. This movement wanted to see ecology as an "alternative", softer and less destructive modern science that could heal nature and people and could be the guide to repairing human–nature relationships (Trepl 1983, Cramer and van den Daele 1985). While there had been hardly any contact points up to then between early conservation and early hygiene – and environmental movements, which aimed at fighting water and air pollution in cities and industrial regions (Ott et al. 1999), this changed now, bringing awareness to both and producing some synergies. Rachel Carson's book is an indication of how environmental pollution (here through pesticides) and conservation (here the disappearance of bird species caused by pesticide use) became linked. Tensions and delimitations, however, persist to this day (see Section 2.4).

Further, important stepping stones to bringing environmental (and conservation) issues, as based on science, into the broader public were the publication of the report of the Club of Rome on the "Limits to growth" (Meadows et al. 1972) and the UN Conference on the Human Environment, which took place in the same year in Stockholm.

Another new stream of thought, which increasingly influenced conservationists and conservation-minded scientists, was the development of environmental ethics and conservation ethics. Although approaches towards such an ethic reach far back in history, they remained quite isolated and only in the 1970s an own field of environmental ethics formed (Brennan and Yeuk-Sze 2015, Gardiner and Thompson 2017). The first specialised journal, named aptly *Environmental Ethics*, was launched in 1979. The debates initiated here brought new ideas of what and why to preserve nature, both to conservationists at large as well as conservation scientists.

With the rise of the environmental movement, with increasing concerns on species losses, growing research in ecology and the development of environmental philosophy, existing conservation concepts (such as wilderness and restoration) were supported and partly reframed by scientific knowledge, as was the management of protected areas. In addition, new conservation approaches were formulated by scientists.

In 1985, Michael Soulé published his seminal paper "What is conservation biology?" and founded the Society for Conservation Biology (SCB). Although there had been earlier books on the theme (e.g. Ehrenfeld 1970), 1985 is mostly considered as the starting point of conservation biology as a discipline. In spite of its name, the proposed discipline from the beginning extended the scientific treatment beyond biology and ecology, also including other disciplines like sociology or "ecophilosophy" (Soulé 1985; see Meine et al. 2006 for more on the history of conservation biology).

A major theme of conservation in the 20th (and 21st) century was concern about the increasing extinction (or perhaps better: eradication) of species (Meine et al. 2006, Adams 2004). It was epitomised by the Red Lists of Threatened Species, since 1964 compiled by IUCN. The theme was brought into the broader public view by books such as *Silent Spring* and later Paul and Anne Ehrlich's book *Extinction* (1981). Against this background, the concept of biodiversity was "invented" in the course of a conference called the "Forum on BioDiversity" which took place in 1986 in Washington DC (see Takacs 1996 and Sections 3.4 and 5.1 for details on the background of the biodiversity concept). The ideas formulated there laid an important basis for the later (1992) CBD. The CBD also established links between conservation, human use of nature, and questions of justice related to it.

Sometime later, (ecological) economists and landscape planners sided with conservation scientists and developed the concept of ecosystem services. This new concept was explained and popularised in two seminal publications in 1997, by Daily (1997) and Costanza et al. (1997). The term had already been used in the Ehrlichs' book of 1981 but in a more restricted meaning, and similar ideas about the usefulness of nature in more than extractive ways had already been conceptualised as "nature's functions" by European landscape ecologists (e.g. de Groot 1992). In the same way as the CBD brought the biodiversity concept into the political arena, the ecosystem services concept became popular by being used as a central organising scheme in the Millennium Ecosystem Assessment (MA) commissioned by the United Nations and published mostly in 2005.

A more recent important step to link conservation, science, and policy was the creation of the IPBES in 2010, as kind of a conservation-focused counterpart to the Intergovernmental Panel on Climate Change (IPCC).

The CBD, the MA, and IPBES all extend conservation beyond the protection of "pristine" or "wild" nature alone and aim at reconciling human use and the protection of nature. The preservation of cultural landscapes and the many cultural dimensions of conservation nevertheless largely remained besides the mainstream of late-20th-century conservation. It was discussed more in the field of heritage protection than in conservation science, though it retained its importance in the practice of conservation in Europe and also other regions, e.g. in Asia (von Droste et al. 1995).

This development of an increasing scientification of conservation and the dissociation from approaches that emphasise symbolic and cultural dimensions of conservation – which dominated its very beginnings – has also raised substantial criticism. The relations between conservation, the natural sciences, and these other dimensions, including the social, the philosophical, and the political, is a theme that I will discuss on several occasions during this book as it is central to any choices about applying conservation concepts in an appropriate way.

The motivations for protecting nature have in fact become even more diverse in the last decades. Many new prudential arguments relating to the need of securing human well-being (based on new insights from the sciences) and even survival have been added. Also, discussions inspired by environmental ethics opened up the spectrum of motivations and arguments, e.g, regarding the proper objects of conservation and the moral justifications for protecting them, both for human purposes but also as ends in themselves. Another issue that environmental ethics raised is to consider the consequences of human activities regarding nature (including conservation) for other people – a matter of intra- and intergenerational justice (see Section 5.2).

To sum up, conservation, as we know it today, originated for the most part in the second half of the 19th century. Driven by several societal developments, such as rapid population growth and industrialisation, a major reason for the beginning of conservation as an organised endeavour was the perception of loss, be it of specific species, remarkable rare natural features, "pristine" landscapes, or cultural landscapes. The specific motivations for conservation were highly diverse from the beginning, mainly derived from symbolic and aesthetic values, ideas of civilisation and people's identity, through utilitarian reasons such as the protection of natural resources. By around 1945, almost all of today's main ideas and motivations for conservation were in place, even though often under different names than today. After 1945, conservation efforts increased and became more and more internationalised, with many international organisations and international agreements being developed. Also, the importance of science increased strongly, not the least with the creation of conservation biology as an own discipline in the 1980s. With it, and with growing public awareness of the need for protecting nature, the complexity of conservation concepts increased. What is still common, and quite obvious, is that conservationists, in one or the other way, all want to protect "nature". But at the same time, many conservationists disagree about what nature *is* or which kind of nature deserves protection, something that also runs through all of the history of conservation.

Thinking about the concept(s) of nature is therefore not just an academic exercise but highly relevant not only for the philosophy of conservation but also for its practice. While this issue will be dealt with in much detail in Chapter 4, it is necessary for the further understanding of this book to discuss it briefly already here.

2.3 What is nature? A preliminary but necessary note

Two books which discuss the meaning of "nature" and the relationship between humans and nature open with almost the same sentence:

> "Nature", as Raymond Williams has remarked, is one of the most complex words in the language.
> *(Soper 1995, p. 1)*

> *Nature* is one of the most complicated terms in English or any language.
> *(Peterson 2001, p. 1)*

In fact, there are multiple meanings of the term "nature". "Nature" and the related adjective "natural", however, are crucial terms. Conservation (in our context) is conservation of *nature*, and the *natural* is often a benchmark when it comes to characterising desired states of things, be it landscapes, objects in specific places, or processes. Heated debates within the conservation community arise from the questions of what is natural and what should thus be preserved or restored: does, e.g., a rural ("cultural") landscape, or even a brownfield or garden qualify as nature and thus as a legitimate object of conservation, or is only "wild" nature "true nature"?

It may appear paradoxically, but in fact "nature" is not a concept of the natural sciences. One will very rarely only find an entry "nature" in dictionaries of biology or ecology, but it is commonly found – and with often long entries – in dictionaries of philosophy and also in those of anthropology and even sociology. Also, a recent textbook on conservation biology (van Dyke and Lamb 2020) neither lists "nature" in its glossary nor in its register. It seems it is taken for granted that biologists, ecologists, or conservation biologists know what "nature" is. But in fact, as will be seen throughout the following chapters, this is far from clear. And the question of "what is nature" has high practical relevance. The much higher attention that the humanities give to the concept of nature stems for the fact that "nature" is (and has always been) a relational term that is defined in relation to humans and their lives. It is often defined in juxtapositions, especially, nature and art (natural vs. artificial), nature and culture, and the natural and the supernatural. There are literally shelves of books and papers about what can or should be understood as nature and natural. The understanding of what *is* nature is also highly variable within cultures and even more across cultures (see Section 4.3); in some cultures there is not even a concept of nature at all.

Stating that nature is not a concept of the natural sciences is not to say that they have no bearing on the concept of nature. Results from, e.g., evolutionary biology or ecology have had high impacts on what we perceive as natural, of how we position ourselves as humans, in or against nature. While there certainly is some material world "out there", which as a whole we may call nature, nature is also to a considerable degree a cultural concept, a social "construction" – but is not a *mere* construction (see Section 4.2).

The important thing to clarify here for the moment is that there is not one "right" concept of nature but that there are many different ideas about *what* nature is and *how* nature is (e.g. robust or fragile). One could even say that there is not one nature but that there are many *natures*, depending on specific cultures, worldviews, and personal relations to the world – as will be seen in detail in the further chapters. These differences in the understanding of what nature is do not only emerge when we compare our "Western" or even scientific view (in the sense of conservation science) with those of other, specifically indigenous cultures. Even within the Western conservation perspective, there is a multitude of highly divergent viewpoints on this issue.

The question of "what nature is" is intimately connected to the question of what it means to be human (Peterson 2001). Defining both what is nature and what is human is at the same time about searching for differences and for commonalities between us (as humans) and the world. We have to put ourselves in contrast to this world at least as an analytical necessity, even when we try to bridge dichotomies between humans and nature. If one says, as often expressed in a conservation context, that the basic fault that leads to environmental destruction and to the extinction of species is already our understanding of humans as being separate from nature, the question remains what it could mean for humans to be "one" with nature. Is it then possible to speak about nature conservation and human responsibility for nature at all? I will come back to these issues in Chapter 4.

2.4 Fuzzy edges: delimiting conservation and conservation-related research

Conservation as practice: related fields

The boundaries of conservation as a field of practice, as well as those of the related research field conservation biology, are not always clear-cut. Conservation grades into several related endeavours, and the question what belongs to conservation and what not has caused considerable controversy (see the case study in Section 2.8).

There are several fields which, although mostly not considered part of conservation, are overlapping with conservation. These are, especially, natural resource management, environmental protection, management for sustainability or sustainable development, and animal protection (animal welfare). On the other hand, major approaches that are often – but not always – considered as part of conservation are preservation and restoration. Conservation is also connected to many other issues (as mirrored already in the field of sustainability), such as poverty reduction and several other societal fields (see Chapter 5). Part of today's conservationists' efforts is also to *mainstream* conservation into other policy fields or into society at large. Many of the aforementioned fields are supported by research disciplines, some of which are of rather recent origin (see below).

Natural resource management overlaps with conservation and, as shown in Section 2.2, is one of the origins and motivations of conservation. Forest management and wildlife management have provided important impulses for conservation, be it with establishing protected area categories or simply by developing practical means to arrive at various conservation goals. As natural resource management is focused clearly on the human use of nature, it is sometimes in conflict with other ideas of nature conservation that emphasise non-use values or even the intrinsic value of nature, i.e. value of nature as independent of any human interests.

The **protection of animals** is also a classic root of conservation. However, as it is focused on the protection of individual organisms and not on that of species (populations), communities, or ecosystems, it easily runs into conflict with most of today's conservation, e.g. when it comes to eradicate invasive species or even when it comes to saving individual animals versus "letting nature take its course" as focusing on natural processes or "the ecosystem" (see Callicott 1980, Rolston 1990, and case study on the Oostvaardersplassen area in Section 3.1). More recently, new efforts attempt to bring together animal protection and conservation under the heading of "compassionate conservation" (see Coghlan and Cardilini 2022).

Environmental protection developed largely independently of conservation during the 19th and early 20th century. It aims mostly at preventing or mitigating effects of the environment on human health and well-being, such as those of air and water pollution or, more recently, climate change. This has been and is reflected often in different institutions. In Germany, e.g., environmental protection and conservation today are both under the responsibility of the Federal Ministry for the Environment, Nature Conservation, Nuclear Safety and Consumer Protection (BMUV), but until the late 1980s environmental protection was under the auspices of the Ministry of the Interior and conservation under that of the Ministry of Agriculture. This is still reflected also in different agencies, with the much older Federal Agency for Nature Conservation (*Bundesamt für Naturschutz*), going back to the early 20th century, working separately from the Environmental Protection Agency (*Umweltbundesamt*) founded in 1974. Between the practice of conservation and environmental protection, the boundaries are also not always clear-cut. The interests and measures of both fields overlap but also are sometimes competing, e.g. when it comes to trade-offs between establishing wind parks versus conservation concerns.

Management for sustainability or sustainable development is a comparatively recent development. Although management for sustainable land and resource use has a long tradition in forestry, today's idea of sustainable management emerged only in the 1980s. Even though there is a plethora of definitions of sustainability today, the general idea still valid today was put forward by the World Commission on Environment and Development, established by the United Nations in 1983. Their report, usually called the Brundtland Report after Gro Harlem Brundtland, then Prime Minister of Norway and chairwoman of the Commission, was published in 1987 (World Commission on Human and Environment Development 1987). Defining sustainable development as "the kind of development which meets the needs of the present without compromising the ability of future generations to meet their own needs" (p. 43), it emphasises three related dimensions of sustainability: the environmental, the social, and the economic dimension. Conservation is thus (included in the environmental dimension) a part of management for sustainability and embedded in this overarching notion. This is also visible in the Sustainable Development Goals (SDGs), where issues of conservation are included in some of the goals, such as SDGs 14 ("Life below water") and 15 ("Life on land").

Preservation and restoration: parts of conservation or distinct fields?

Discussions about the differences between conservation and preservation have their origin in the USA, especially during the early 20th century (see Section 2.2). The classical distinction here is between *preservation* as protecting (parts of) nature from any human influence (especially in the context of maintaining wilderness) and *conservation* as protecting areas and objects which at the same time are (wisely and sustainably) used by humans (e.g. Norton 1986, van Dyke and Lamb 2020). It was further complicated by the common attribution that preservation was linked to a non-anthropocentric view of nature and conservation to an anthropocentric, completely utilitarian view. Already Norton (1986), however, demonstrated that this is a problematic linkage and can neither be validated historically (e.g. with the classical opposition between the approaches of John Muir and Gifford Pinchot) nor philosophically. Today, the term "preservation" (with respect to nature) is mostly used in the compound expression "wilderness preservation" and implies refraining from human intervention in "pristine" areas. In the current usage of "conservation", however, the concept mostly includes also ideas of preservation and is not focused exclusively on a utilitarian perspective. For the purpose of the book, I will consider preservation as part of conservation, the more as it fits very well under the broad definition of conservation given earlier. Also *intentionally* refraining from interfering with nature ("let nature take its course") is a human action, one that is part of the instruments of nature conservation.

The same holds for *restoration*. I consider restoring nature also as a conservation activity, even though sometimes the research fields devoted to them are seen as separate disciplines (restoration ecology vs. conservation biology). Both fields serve the same ends, namely to protect non-human nature, either by maintaining its current state or by bringing altered areas into a desired state again. Wiens and Hobbs (2015) provide a good discussion of the differences as well as the overlap between conservation and restoration; the authors also include classic "preservation" as part of conservation.

So, preservation and restoration are all dealt with here under the umbrella of conservation. I will follow the same approach. This does not mean that there are no differences between the approaches (it is self-evident that there are), but for the purpose of this book, a broad scope of "conservation" and "conservation concepts" is appropriate.

Research disciplines and research communities linked to nature conservation

The discussion about conservation takes place in different research communities, which only partly overlap. As mentioned earlier, conservation biology itself was founded as a distinct discipline, with its own society and own journal, only in the mid-1980s. But there are several other disciplines which either contribute to conservation biology or overlap with it. Soulé, in his 1985 landmark paper, includes several biological disciplines as part of or contributing to conservation biology, e.g. population biology, genetics, physiology, biogeography, but also natural resource management, sociology, and ecophilosophy (see also left part of Figure 2.7 in Section 2.8). Soulé emphasises that conservation biology is a "synthetic, multidisciplinary" science that crosses "the borders between disciplines and between 'basic' and 'applied' research" (Soulé 1985, p. 728). As described in the case study in Section 2.8, there have been debates about extending "conservation biology" into "conservation science". Unless specifically indicated, however, I will use both terms synonymously.

Next to discussions within conservation biology "proper", important discourses on conservation take place in political ecology, but also within environmental philosophy (especially environmental ethics) as well cultural ecology and ecological anthropology. As to my experience, conservation biologists are often not much aware of these other research fields. Although at least environmental ethics has become more known in conservation circles (academic and non-academic), these fields still have rather distinct research communities. As they provide some interesting perspectives on conservation and can contribute very important contexts to conservation, they will be described here in brief.

Human ecology as a discipline developed in the 1920s as an effort to investigate the interrelations between humans, their societies and culture, and their (biophysical) environment. The origins of the interdisciplinary field are in sociology, anthropology, and geography. In the course of its history it has developed in various directions and into various subfields (Sutton and Anderson 2013, Knapp 2017). One of these fields is **cultural ecology** (since around the 1940s), its special emphasis being on the adaptation between humans and their respective environment through cultural means, and often with a strong historical and place-based component. Today's cultural ecology is neither pursuing the idea of an environmental determinism (i.e. culture follows directly from a specific environment) nor the opposite approach that culture has the decisive role in shaping humans' interactions with nature.

Instead it focuses on a dynamic interaction between nature and culture, increasingly understood as unity instead of a dichotomy (Teherani-Krönner and Glaeser 2020). Obviously, through this focus on human–nature relationships, cultural ecology also has a high relevance to conservation issues and their background (see Sutton and Anderson 2013).

A more recent but very important offspring from human and cultural ecology is **political ecology**. Political *ecology* (as well as cultural ecology) is of course not ecology or a branch of the same in the original sense of the natural sciences (such as animal ecology, community ecology, or physiological ecology). It is an interdisciplinary field that investigates environmental change and the relations between humans and nature associated to it, but the starting point here is less nature but humans and their complex societies. Going beyond most of cultural ecology, political ecology especially emphasises economic structures and power relations and the ideologies related to them. It originated in the 1970s and 1980s from scholars of different disciplines, most importantly human geography (especially development geography) and anthropology (Neumann 2005, Roberts 2020).

Robbins (2012) sees political ecology not as a clear-cut discipline but more as a highly diverse "community of practice" united by a common theme, which he describes in the following way:

> [B]roadly they can be understood to address the condition and change of social/environmental systems, with explicit consideration of relations of power.
> *(Robbins 2012, p. 20)*

Classical question posed by researchers are therefore:

> Whose use of, claims to, and/or perceptions of the environment prevail, and why?
> *(Karlsson 2015, quoted in Roberts 2020, p. 2)*

They aim at understanding environmental conflicts and their relation to environmental degradation and social marginalisation. This involves considering various social factors, especially the forces of political economy – on various spatial scales, from local to global. Quite obviously it also necessitates to understand the specific ecological conditions and their dynamics. Striking the balance between the social sciences and the natural sciences has been and is an ongoing challenge within political ecology (Roberts 2020).

Political ecologists often emphasise that even "unpolitical ecology" (or better "unpolitical conservation and conservation biology") is always implicitly political. As described earlier, in the history of conservation, conservation measures, especially the establishment of protected areas, are highly political, because conservation, as Adams (2015, p. 65) states, "involves making choices about the relations between people and nature".

While political ecology is often very critical about classical conservation, especially in developing countries, it is also searching for solutions to the problems it describes. A recent contribution that demonstrates the high relevance of political ecology for conservation is the book by Büscher and Fletcher (2020), which not only uses an approach from political ecology for a critique of ongoing conservation tendencies but also proposes an alternative to these ("convivial conservation"; see Section 5.2).

2.5 What are conservation concepts?

Building on the earlier definition of conservation, conservation concepts *are concepts that were developed for and/or are used to promote and implement conservation.*

This is again a very broad definition, and it must be clear from the outset that conservation concepts are of very different kinds, ranging from merely technical concepts developed in conservation biology, via concepts referring to the objects of conservation, through complex concepts that stand for a full-fledged approach of conservation, such as ecosystem services or ecosystem management. Thus, not all concepts are created equal.

Concepts should not be confused with *terms*. The same term (e.g. "ecosystem" or "community") may denote different concepts, with different definitions and meanings, and the same concept may be designated by different terms. The discussion in this book is mainly not about terms and terminology but about concepts and the ideas behind them. Concepts may be characterised by specific formal definitions, but in the end it is the use of words that determines their meaning (Schwarz 2011). These uses can change over time, and with it the specific meaning of terms. Concepts are highly dynamic and while this sometimes leads to problems in communication about the proper content of concepts, it is often just this flexibility, adjustment, and "continual reprocessing" (Schwarz 2011, p. 20) that makes a concept useful (see Section 3.4). Please note also that concepts in this sense are not true or false but only appropriate or inappropriate to specific tasks at hand.

The question who determines the (dominant) meaning of a concept, however, is not only a scientific one. Concepts – and even knowledge about nature as such – are not neutral, they often carry with them various values and even political programmes. Conservation concepts therefore can also have high normative power (Jax et al. 2013).

Sometimes, also the selection of *terms* and *language* is important. More than in many other fields, conservation uses a mix of everyday terms, technical terms, and terms that were specifically developed for addressing specific audiences outside of the inner circle of conservation scientists – in order to increase motivation for conservation; examples are ecosystem services and even biodiversity (see Sections 3.4 and 5.1). Especially on behalf of the latter purpose, terminology and the delimitation of concepts' meanings lead to considerable controversies within conservation biology. Beyond that, language is often a matter of power. To quote geographer and historian David Livingstone:

> To have command of definition is to have control of discourse.[...] The adjudication of definitions is, of course, an inherent boundary-marking, or boundary-making, enterprise designed to demarcate the true from the false, the legitimate from the illegitimate, the relevant from the irrelevant. Accordingly, the ownership of terminology is of enormous consequence in dialogue, for by it both ideas and people can be positioned on particular sides of debates. To dictate definition is to wield cultural power.
> *(Livingstone 1992, quoted in Jones 2003, p. 24)*

This still has an enormous importance in conservation, as will be seen when it, e.g., comes to agreements on a common vocabulary for conservation within but also beyond "Western science" (see case study in Section 4.4 on the IPBES conceptual framework). Imposing particular terminologies and concepts for conservation is sometimes even perceived as a continuation of Western colonialism. This pertains to both to seemingly "neutral" scientific terms as well as ideas of nature and of progress imbedded in them (see. Ross 2015 and Section 5.2).

2.6 Conservation concepts and their societal contexts

Conservation and its concepts cannot be understood outside their social contexts. Conservation is about how we want to live from, in, and with nature (O'Neill et al. 2008, Pascual et al. 2022). Conservation concepts derive from people; they affect people; they have to be balanced with other interests of people; they have to be implemented by and with people. Also, conservation biologists are a scientific community composed of different persons with strongly diverging goals and interests – even though it may sometimes appear as if the specific goals of (biological) conservation were obvious. They are not.

Conservation biology as a normative science

Conservation and conservation biology (as research on and for conservation) are often intertwined and have influenced each other. As the history of conservation shows, conservation ideas preceded conservation biology, but research has also always influenced the direction that conservation has taken. Since a couple of decades these mutual influences between conservation, conservation biology, and conservation policy have become stronger and stronger to a point where it is difficult to entangle the mutual lines of influence. This has also consequences for the very nature of conservation biology and conservation concepts.

Conservation biology, already in the way Soulé (1985) defined it, is not neutral, not merely descriptive, and it cannot be so. It is based on norms and is replete with what Williams (1985) called morally "thick" concepts, meaning concepts that are not only descriptive but also evaluative, normative. In everyday life, such a "thick" concept is, e.g., "friendship", which not only describes a relation between persons but also has a positive connotation as something morally good or even desirable. "Conservation" is in itself is already such a "thick" concept, formulating a specific societal goal (something desirable) in terms of supporting a specific direction of human actions, i.e. nature conservation. Research in conservation biology is oriented towards supporting these goals. Likewise "biodiversity" is not a merely descriptive term but is considered in itself as something "good", and thus is also normative, as something which humans should strive to maintain or increase – as are ecosystem services, biological integrity, ecosystem health, and many other concepts (Callicott et al. 1999). There are many epistemic-moral hybrids, as Potthast (2000) called them, in which the descriptive is mixed with the normative, so they serve as terms to describe something but also to express value. That does not mean that all terms in conservation biology are normative; there are many technical terms with little or no value dimensions at all; examples would be "critical habitat", "edge effect", or "population viability analysis". But the whole discipline of conservation biology takes place in a normative setting. Given this setting, it is sometimes difficult to separate descriptive and normative dimensions – and uses – of conservation concepts. In real life both dimensions are almost inseparable. However, at least analytically, a distinction between the normative and the descriptive dimensions remains a desiderate for the sake of clearer arguments (see also Section 2.7 about the roles of science and its "objectivity" in a conservation context).

Is there an overarching "we" in conservation?

Up to now, I mostly talked about "society" as some fixed entity. Both "society" and "humanity", and also "the conservationists", are, however, not fixed and homogeneous entities. In fact, they consist of many different groups with different ideas and interests, groups which also have different access to power and voice in conservation discourses. These differences have to be accounted for, because conservation is as much about people as it is about nature. For example, the very concept of nature, is not given "as such" or by science, and thus has different meanings for different groups of people (see Section 2.3 and Chapter 4). Although the need for conservation, or even particular directions of the same, may appear as evident, different perspectives exist, within the community of conservationists, but even more when one leaves the Western realm. Thus, there is no simple overarching "we" in conservation.

This relates also to juxtaposing the "Western" and "non-Western". "Western" and the "West" are here used not as a geographical but as cultural terms. What is perceived as conservation and conservation science today, as evidenced in its journals, societies, and NGOs, originated in Europe and the North America, as described earlier. These cultures, characterised by the Enlightenment, ideas of progress, the increasing importance of science, and a complex set of norms and values, have spread (and in part have been spread forcefully) far across the globe. This happened at first with the colonial powers but, more recently, also through science and technology and through the dominant economic system. Most conservationists grew up in this Western culture and, as in any culture, most of those who live in a culture take many things for granted. They sometimes also consider their own set of ideas, values, and practices as being superior to those of other cultures. Western culture (however defined in detail) has prompted this apparent superiority not the least by the successes of its scientific methods and the technologies enabled by them. But "we" (explicitly including myself), as persons socialised in "the West", are sometimes immersed so deeply into our received culture and so convinced of the method of science, etc. that we are in danger of being blind to the fact that our perspective is only one among many others, and that many things, in fact, cannot be taken for granted. Our "we" cannot be simply equated with "humanity" and many insights from Western culture not considered to be valid for all humans. The margins of "Western culture" and Western thinking are not sharp; as in any society, in any culture, there is diversity, even within "the" community of scientists or conservationists. There are people coming in "from outside" and/or developing ideas strongly deviating from the mainstream. So even here, there is no common "we". Nevertheless I will speak of "we" now and then, e.g. as a coarse description of what I sketched as "Western" culture, as "we" denoting "humanity" and "we" as meaning all conservationists. I will try, however, to differentiate the broad "we" whenever necessary.

The same holds for "non-Western" cultures. They are highly diverse and cannot simply be juxtaposed as one bloc against the Western cultures (see Sections 4.3 and 4.4). Complete homogeneity exists in no culture, in no society. In most societies there are people (or groups of people) with different knowledge, different power, different esteem, and different wealth; there are different genders and different biological characteristics. All this should be kept in mind in connection with discussing conservation and good human relationships with nature (see especially Chapter 5).

2.7 Science and conservation: an ambiguous relation

As shown in the section on the history of conservation (Section 2.2), most of conservation started without much science or without science at all. Experiences of loss of natural features were connected with motivations that emphasised aesthetics, care of animals, ideas of heritage, symbolic aspects of nature such as individual and national identity, but also utilitarian aspects. The systematic devotion of science towards conservation issues – in the form of ecology and conservation biology – is a rather late development. The mutual relations between science and conservation have remained ambiguous to this day, for a couple of reasons and both from the side of conservation and from that of science.

Form the side of science, I briefly described some tensions in the context of the history of conservation discussed earlier, e.g. for the Ecological Society of America. The point here is mostly if and to what degree science, here specifically ecology, should be engaged in conservation issues, on behalf of its own self-limitation as a neutral and "objective" discipline and its public reputation following from this "objectivity". In the early 2000s there were debates in the Ecological Society of Germany, Austria and Switzerland (*Gesellschaft für Ökologie*, GfÖ) about the degree to which the society should be perceived publicly as a real *scientific* society and not so much as a conservation-oriented advocacy institution. The issue of advocacy and the different roles of scientists in environmental issues (but not only there) is a matter of discussion to this day (Pielke 2007, Büttner et al. 2023). Science can take different roles in the context of conservation (Jax 2003). Its empirical, sometimes even prognostic, role is undisputed. Scientists can do empirical research within a set of given goals without compromising the classical scientific standards of conduct; science in this manner is necessary and has strongly improved conservation (as does research for human health). Not the least, it has allowed a much more precise description of what is lost (or in danger of being lost) and why. Also, there is a theoretical-heuristic role of science. Connecting both classical scientific standards and epistemological standards, science here serves the identification of knowledge gaps and uncertainties, pushing for the clarification of concepts, questions, and goals (so that they may become more operational for research and management); it finally can help in scrutinising the consistency of means and ends for practical conservation management. Ecological theory itself is highly important for contextualising empirical data; dominant theories can be contested or change over time (e.g. from the "balance of nature" to "the flux of nature", Pickett et al. 1992), requiring new interpretations of empirical results, with potential consequences for conservation management. The sometimes postulated role of ecology as normative and goal-setting, i.e. justifying and directing conservation goals, however, is not something ecology could provide without losing its self-defined character as a natural science. Values and norms cannot be taken directly from nature, from empirical data.

In part linked to the question of the roles that science can play for conservation, the other side of the debate asks how much science really can and should contribute to conservation. While nobody denies the importance of science and its necessary contributions to conservation, there are substantial concerns that science (and economics) is nowadays dominating conservation too much. The scientific analysis and characterisation of conservation objects (e.g. forest, landscape) brings explanation and guidance for action, but also problems. A "scientification" of conservation may lead to missing its real target objects. These target objects are mostly not specific scientifically definable objects such as the "ecosystem", but much more beauty, sublimity, symbolic dimensions of nature, and various personal relations to nature – echoing the origins of most conservation efforts (see earlier and Sagoff 2013, Kirchhoff 2019).

By the same token, an overly emphasis on Western science, as taken for granted by most conservationists, can impede the linkage between different knowledge forms for conservation. Non-Western concepts of nature (e.g. ideas of "mother nature") are often at odds with the established science-based concepts and cannot be simply "translated" into scientific language (see Chapter 4).

2.8 Case study: the debate about the "new conservation"

To illustrate the relevance of many issues discussed in this chapter, a case study will be helpful. The debate about what constitutes "true" conservation, which took place in the 2010s, can serve as a good example here.

In 1985, Michael Soulé published a landmark paper for conservation biology. "What is conservation biology?" is seen as kind of the founding manifest of the discipline. In this paper, Soulé laid out what he saw as the major principles of the new discipline, in the form of both "functional postulates" and "normative postulates". Twenty-seven years later, Peter Kareiva and Michelle Marvier published a paper in the same journal (*BioScience*) entitled "What is conservation science?". In that paper, explicitly mirroring in its structure with Soulé's older paper, they challenged the "old conservation" and described what they saw as a necessary contemporary "update" of conservation biology as "new conservation". This new conservation should overcome the "flaws" of traditional conservation. The publication started an intense, in part highly emotional, debate about the question what "true conservation" (e.g. Marvier 2014) is. I will use this debate to demonstrate some directions of major dissent *within* the conservation community. What will become evident is that dissent in conservation biology is mostly not a matter of scientific or technical issues but one of values and choices. Also, I will demonstrate that this and similar debates are often building on highly exaggerated descriptions of the respective opponents' positions, which can distract from broader and more united conservation efforts.

So what reasons were given for the need to "update" conservation biology, what are the main differences between the "old" and the "new" conservation? One critique of "classical" conservation biology is that it was too much focused on the biological sciences. In their paper, Kareiva and Marvier (2012; in the following K&M) present a graph (Figure 2.7) that is meant to show differences between the old conservation *biology* and new conservation *science* with respect to the fields contributing to both; they understand conservation science as "a broader and more interdisciplinary endeavor to protect nature" (p. 963).

In Kareiva and Marvier's scheme, Soulé's original depiction of disciplines contributing to conservation *biology*, taken from Soulé (1985), is presented as part a (left).[6] In depicting conservation *science* (part b, right), Soulé's conservation biology is only one segment of the broader approach. While most of Soulé's fields contributing to conservation biology are indeed from the biological sciences, there are, however, important exceptions. These are the "social sciences" and

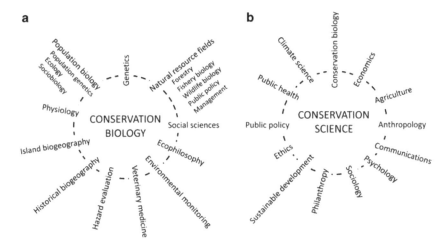

FIGURE 2.7 Fields contributing to conservation biology (a) and those contributing to conservation science (b).

From Kareiva, P. and Marvier, M. (2012). What is conservation science? BioScience, 62, 962–969; reprinted by permission of Oxford University Press.

"ecophilosophy" and also some applied, "natural resource" fields, including "public policy" (as visible in part a of the graph). In contrast, the fields in K&M's depiction of conservation *science* are mostly taken from the humanities. In addition they added some new applied endeavours such as agriculture, public health, and even philanthropy. But what happened is that K&M, in their scheme of conservation science, have taken out Soulé's "social sciences" and "ecophilosophy" and also his applied fields from "conservation biology". Instead, for their "conservation science", they list them separately, which leaves only the biological or biophysical sciences as constituting "conservation biology" – which was never Soulé's idea.

The main critique from Kareiva, Marvier, and other "new conservationists", related to the claimed narrowness of traditional conservation biology, is that conservation biology is "concerned solely with the welfare of non-human nature" while conservation science "has as a key goal the improvement of human well-being through the management of the environment" and at the same time aims to "jointly maximize benefits to people and to biodiversity" (all quotes from K&M, p. 962). This goes along with the allegation that conservation biology was perceiving of people only either in the role of a threat to nature or, – meaning mainly Western conservationists and scientists – as protectors and saviours of nature. K&M call for a "richer set of roles for people in conservation" (ibid.).

To corroborate and extend this critique, K&M take up the Soulé's postulates for conservation biology and juxtapose them with their own postulates for conservation science. Soulé's four "functional postulates" remain within ecological and evolutionary science. Postulate 1, e.g., reads: "Many of the species that constitute natural communities are the product of coevolutionary processes" (Soulé 1985, p. 729). Soulé's "normative postulates" then are explicitly strongly normative in a moral sense and postulate the value of several phenomena of nature, such as "Evolution is good" or "Biotic diversity has intrinsic value" (ibid., p. 731). For Soulé, these postulates are meant to "provide standards by which our actions can be measured" (ibid., p. 730). The functional postulates of Kareiva and Marvier already differ from Soulé's postulates in that they refer partly to ecological, partly to political and social conditions, e.g. "Nature can be surprisingly resilient" (K&M, p. 965) or "local conservation efforts are deeply connected to global forces" (ibid., p. 966). The normative postulates of K&M, however, are of a completely different character than those of Soulé. They are not direct statements about (moral) values, but "instead, offer practical statements of what conservation should do in order to succeed" (ibid.). Examples are "Conservation must occur within human altered landscapes" (ibid.) or "Conservationists must work with corporations" (p. 967). I think it was a missed chance here that K&M did not – at least in part of their normative postulates – respond to Soulé with clear value statements of their own but chose a completely different category of normative postulates.

Be that as it may, the paper of Kareiva and Marvier spawned an intense debate, with responses directly from Soulé and also from many other authors (e.g. Noss et al. 2013, Doak et al. 2014, Greenwald et al. 2013, Tallis and Lubchenco 2014) and rejoinders by Kareiva and Marvier (e.g. Marvier 2014, Kareiva 2014). All participants saw the debate as important for the future direction of conservation.

Besides the question what is really "new" in the "new conservation" (Greenwald et al. 2013, Meine 2014), the controversy revolves around a few major points. The most important issue could be phrased by the question: *What is conservation for? – for human needs and well-being or for the sake of nature (or biodiversity) as such*? In fact it is not about either-or but about what takes priority. "New conservationists" complain inter alia about conservation neglecting the needs of disadvantaged peoples, e.g. in developing countries, and postulate a stronger emphasis on protecting those parts of nature which are useful to people. This is linked with question of what *values* nature has: only utilitarian ones, instrumental ones (to further human well-being), intrinsic ones (for the sake of nature itself) – or yet others (see Section 3.1 for a discussion of value types)?

Interestingly, some *opponents* of the new conservation complain that "[t]he priorities of NCS [new conservation science] rest on ethical values, not science" (Doak et al. 2014, p. 79). But one might ask what else could they rest on? Also "old" conservation is based on values. Conservation is not just applied science, as also Soulé's normative postulates show as well as the emphasis of much of traditional conservation on the intrinsic value of nature. This has often been expressed as a dichotomy between two ethical positions, an "anthropocentric" and an "ecocentric" one – a dichotomy which, as I will argue later, is neither helpful for conservation nor does it reflect the actual breadth of values related to nature.

The question is more on how different values, also beyond that seeming dichotomy, *are to be balanced*. The argument of Doak and colleagues may be seen in the context of their warning that the "new conservation" is not based enough on "facts" and ecological knowledge (see also Soulé 2013); the same allegation, however, is made in turn by Kareiva and Marvier (2012) with respect to the "old" conservation, namely that the old conservation was not evidence-based.

Situated halfway between ethical and strategic considerations is the moot question of *the role of the economy* in conservation, and more specifically, of possible cooperations with large corporations. This issue has a prominent place in K&M's argumentation. Including economic aspects is seen as a moral duty, in terms of accounting for utilitarian values of nature (e.g. of its ecosystem services) for the well-being of people, especially for disadvantaged people. But equally important appears to be the *strategic* aspect of including economic issues: to further motivation for conservation among more people and to win new powerful allies for conservation through working with corporations. K&M acknowledge the danger of greenwashing but still see more benefits than problems here, if handled carefully. For Soulé and others, however, this amounts to giving up on the very aims of conservation. Soulé (2013) states that new conservation "promotes economic development, poverty alleviation, and corporate partnership *as surrogates or substitutes* for endangered species lists, protected areas and other mainstream conservation tools" (p. 895, my emphasis). In addition, many conservationists expressed doubt if the strategic impulses based on economic reasoning will do the job promised by the new conservation.

Another juxtaposition made in the debate, related to a potentially increased economic emphasis, is the question of *what places are worth protecting*. Would the new conservation lead to terminating the protection of wilderness areas and national parks (as Soulé fears), with the focus being mainly laid on "productive landscapes" where people live, such as urban areas and "working landscapes"?

Last but not least, also some "ecological facts" are contested, which both sides, being educated in a scientific tradition, emphasise. Dissent here is about *as to whether nature is robust or fragile*. "Robust" here alludes, e.g., to the functional postulate, by K&M (p. 965), that "nature can be surprisingly resilient", while "old conservation" tends to emphasise the fragility of nature, as one argument for protecting large wild areas.

What becomes visible from the debate about the "new conservation" is that most of the arguments are not about "facts" (be they scientific or social) but about values and about the balance between human needs and those of non-human beings. Even the question if nature is resilient or not is not a purely scientific question but depends in part on choices: what is meant exactly by robust, fragile, or resilient, which part of nature is in focus, and what are the criteria to decide if the (eco) system is still functioning (see Section 3.4 and Jax 2010)?

A second observation is that, given the high personal commitment of most conservationists, the debate is very heated and often involves misreading each other's claims and arguments. This is a broader problem within the conservation community, not limited to this specific example. As Meine (2014) demonstrates quite well, "new conservationists" neglect much of the complexity within traditional conservation, e.g. as related to the alleged neglect of humans and the social in conservation. The concluding claim of K&M that only (?) new conservation is "advocating conservation *for* people rather than *from* people" (p. 968) is likewise an unwarranted exaggeration. Also, the way in which K&M took all the social and applied aspects out of "conservation biology" in the depiction of their new "conservation science" (Figure 2.7) constitutes a distortion. On the other side, Soulé builds a strawmen when he states, as cited earlier, that new conservation is meant to *substitute* protected areas and other traditional conservation approaches. Likewise, Butler (2014, p. x) assembles several arguments he sees as being put forward by the "neo-greens" ("new conservationists" and similar villains…), which, taken together, amount almost to caricature of the "new conservation". To make things worse, both sides lump together many different people with quite distinct ideas and approaches to conservation. Thus, concepts such as ecosystem services, novel ecosystems, the Anthropocene, technocracy ("engineered global garden") are all thrown into the same pot, as being favoured by the "neo-greens" (e.g. in some contributions in Wuerthner et al. 2014). On the other side of the trench, wilderness, the wild, protected areas, and biodiversity are lumped as characteristic of "traditional conservation", even though they do not necessarily protect the same things. Consequently, Meine (2014, p. 45f) has condensed the exaggerated claims of both sides into two "stories" of the old and new conservation ("Once there was…").

In fact, the conservation landscape, i.e. the spectrum of conservation concepts and approaches, is much more complex than can be expressed by a dichotomy as simple as that between an "old" and a "new" conservation. There have been several attempts to differentiate or harmonise the positions described earlier, of which I will name just a few. Both Michelle Marvier and Peter Kareiva wrote respectful rejoinders to Soulé's (2013) critique of their 2012 paper. They tried to disperse misimpressions and searched for common ground. Marvier (2014) argued that the new conservation is not to replace the "old" one but to supplement it, broadening the toolbox. Kareiva (2014) asserts the same for himself and his colleagues from the Nature Conservancy, also stressing that their focus is not economic growth instead of protecting species. Acknowledging this, Soulé (2014) eventually concedes some common ground with the ideas of the new conservation. But divergences remain:

> I believe that some other issues distinguishing the views of Kareiva and myself rest on nearby irreconcilable, beliefs and ideologies not amenable to testing by empirical science. One of these beliefs is the notion that wild things and places have incalculable intrinsic value, at least as salient as the value of humanity.
>
> *(Soulé 2014, p. 637)*

It is neither necessary nor possible to completely integrate all conservation perspectives. There *are* differences. But they are not between two dichotomous positions only. More recently, e.g., Büscher and Fletcher (2020) demonstrated that the lines between the "camps" of conservationist are not that simple and that more differentiation is needed. That is also a major aim of this book: to visualise the wide spectrum of conservation concepts, show where and how they differ – in order to avoid too much unproductive infighting between conservationists. All of the controversial issues raised in this case study will be elaborated in the chapters that follow.

Notes

1

> Vogelschutz ist nicht nur eine Liebhaberei, eine aus ethischen und ästhetischen Motiven hervorgegangene Passion – also nicht nur aus der Bewunderung für der Vögel Gesang, aus dem Bestreben nach Verschönerung und Belebung der Natur hervorgegangen – sondern Vogelschutz ist in erster Linie lediglich eine nationalökonomische Frage, und zwar eine Frage von eminentester Bedeutung.
>
> Vogelschutz ist eine Maßnahme, aus der uns Menschen ein materieller, ein großer pekuniärer Nutzen erwächst. Vogelschutz will die dem Menschen nützlichen, ja direkt notwendigen Vögel schützen und vermehren.

2 According to Lenzing (2003), the term was first used in 1802 in a novel of the French poet Francois-Rene Vicomte de Chateaubriand (1768–1848), as "monument de la nature". Alexander von Humboldt some years later then described an old tree as a natural monument (*Naturdenkmal*).
3 www.nps.gov/yell/learn/management/yellowstoneprotectionact1872.htm. Accessed April 12, 2023.
4 "'Heimat' suggerierte die menschliche Umwelt als natürliche Ordnung und deklarierte das ländliche Leben zum natürlichen Leben" (Schmoll 2004, p. 388).
5 "…die Leere mancher Stunde des langen Winters mit dem Ausdreschen seines Korns auszufüllen, wobei die Energie seiner Muskeln frisch erhalten wird, als vor langer Weile nach der nächsten Eisenbahnstation zu troddeln, um städtische Vergnügungen aufzusuchen."
6 In Soulé's paper from 1995, what is here part (a) of the figure, is juxtaposed with "cancer biology", to demonstrate the synthetic, multidisciplinary character of conservation biology, both being disciplines with basic as well as applied parts.

References

Adams, W.M. (2004). *Against extinction: the story of conservation*. Earthscan, London, New York.
Adams, W.M. (2015). The political ecology of conservation conflicts. In: *Conflicts in conservation: navigating towards solutions* (eds. Redpath, S.M., Gutiérrez, R.J., Wood, K.A. and Young, J.C.). Cambridge University Press, Cambridge, pp. 64–78.
Agrawal, A. (1997). The politics of development and conservation: legacies of colonialism. *Peace & Change*, 22, 463–482.
Allee, W.C., Emerson, A.E., Park, O. et al. (1949). *Principles of animal ecology*. Saunders, Philadelphia.

Berlepsch, H.v. (1899). *Der gesamte Vogelschutz, seine Begründung und Ausführung.* Verlag Eugen Köhler, Gera.

Bildstein, K.L. (2001). Raptors as vermin: A history of human attitudes towards Pennsylvan birds of prey. *Endangered Species Update*, 18, 124–128.

Bocking, S. (1993). Conserving nature and building a science: British ecologists and the origins of the Nature Conservancy. In: *Science and nature: essays in the history of the environmental sciences* (ed. Shortland, M.). Alden Press, Oxford, pp. 89–114.

Bocking, S. (2020). Science and conservation: a history of natural and political landscapes. *Environmental Science & Policy*, 113, 1–6.

Brennan, A. and Lo, Y.-S. (2015). Environmental ethics. In: *The Stanford encyclopedia of philosophy (fall 2015 edition)* (ed. Zalta, E.N.): https://plato.stanford.edu/entries/ethics-environmental/

Brockington, D. and Igoe, J. (2006). Eviction for conservation: a global overview. *Conservation and Society*, 4, 424–470.

Büscher, B. and Fletcher, R. (2020). *The conservation revolution: radical ideas for saving nature beyond the Anthropocene.* Verso, London, New York.

Butler, T. (2014). Introduction: lives not our own. In: *Keeping the wild. Against the domestication of the earth* (eds. Wuerthner, G., Crist, E. and Butler, T.). Island Press, Washington D.C., pp. ix–xv.

Büttner, L., Darbi, M., Haase, A. et al. (2023). Science under pressure. How research is being challenged by the 2030 Agenda. *Sustainability Science*, 18, 1569–1574

Buxton, E.N. (1902). *Two African trips: with notes and suggestions on big game preservation in Africa.* Edward Stanford, London.

Callicott, J.B. (1980). Animal liberation: a triangular affair. *Environmental Ethics*, 2, 311–338.

Callicott, J.B., Crowder, L.B. and Mumford, K. (1999). Current normative concepts in conservation. *Conservation Biology*, 13, 22–35.

Carson, R. (1962). *Silent spring.* Houghton Mifflin, Boston.

Coghlan, S. and Cardilini, A.P.A. (2022). A critical review of the compassionate conservation debate. *Conservation Biology*, 36, e13760.

Conwentz, H. (1904). *Die Gefährdung der Naturdenkmäler und Vorschläge zu ihrer Erhaltung.* Gebrüder Borntraeger, Berlin.

Conwentz, H. (1914). On national and international protection of nature. *Journal of Ecology*, 2, 109–122.

Costanza, R., d´Arge, R., de Groot, R. et al. (1997). The value of the world´s ecosystem services and natural capital. *Nature*, 387, 253–260.

Cramer, J. and van den Daele, W. (1985). Is ecology an "alternative" natural science? *Synthese*, 65, 347–375.

Curry-Lindahl, K. (1978). Background and development of international conservation organizations and their role in the future. *Environmental Conservation*, 5, 163–169.

Daily, G.C. (ed.) (1997). *Nature's services. Societal dependence on natural ecosystems.* Island Press, Washington D.C.

Davis, D.K. (2012). Enclosing nature in North Africa: national parks and the politics of environmental history. In: *Environmental history of the Middle East and North Africa* (ed. Mikhail, A.). Oxford University Press, Oxford, pp. 159–179.

Davis, J.M. (2015). The history of animal protection in the United States. *The American Historian*, www.oah.org/tah/issues/2015/november/the-history-of-animal-protection-in-the-united-states/

de Groot, R.S. (1992). *Functions of nature. Evaluation of nature in environmental planning, management and decision making.* Wolters-Noordhoff, Groningen.

Doak, D.F., Bakker, V.J., Goldstein, B.E. et al. (2014). What is the future of conservation? *Trends in Ecology and Evolution*, 29, 77–81.

Dudley, N. (2013). *Guidelines for applying protected area management categories including IUCN WCPA best practice guidance on recognising protected areas and assigning management categories and governance types.* IUCN, Gland.

Ehrenfeld, D. (1970). *Biological conservation.* Holt Rinehart and Winston, Toronto, Montreal.

Ehrlich, P. and Ehrlich, A. (1981). *Extinction. The causes and consequences of the disappearance of species.* Random House, New York.

Evans, D. (1997). *A history of nature conservation in Britain.* 2nd ed. Routledge, London, New York.

Ford, C. (2004). Nature, culture and conservation in France and her colonies 1840–1940. *Past & Present*, 183, 173–198.

Gardiner, S.M. and Thompson, A. (eds.) (2017). *The Oxford handbook of environmental ethics.* Oxford University Press, Oxford.

Gissibl, B. (2012). A Bavarian Serengeti: space, race and time in the entangled history of nature conservation in East Africa and Germany. In: *Civilizing nature. National parks in global historical perspective* (eds. Gissibl, B., Höhler, S. and Kupper, P.). Berghahn Books, New York, pp. 102–119.

Gissibl, B., Höhler, S. and Kupper, P. (eds.) (2012a). *Civilizing nature. National parks in global historical perspective.* Berghahn Books, New York.

Gissibl, B., Höhler, S. and Kupper, P. (2012b). Introduction. Towards a global history of national parks. In: *Civilizing nature. National parks in global historical perspective* (eds. Gissibl, B., Höhler, S. and Kupper, P.). Berghahn Books, New York, pp. 1–27.

Greenwald, N., Dellasala, D.A. and Terborgh, J.W. (2013). Nothing new in Kareiva and Marvier. *BioScience*, 63, 241.

Grove, R.H. (1990). Colonial conservation, ecological hegemony and popular resistance. In: *Imperialism and the natural world* (ed. MacKenzie, J.M.). Manchester University Press, Manchester, pp. 15–50.

Grove, R.H. (1992). Origins of Western environmentalism. *Scientific American*, 267, 42–47.

Grove, R.H. (1995). *Green imperialism: colonial expansion, tropical island Edens and the origins of environmentalism, 1600–1860.* Cambridge University Press, Cambridge.

Haila, Y. (2012). Genealogy of nature conservation: a political perspective. *Nature Conservation*, 1, 27–52.

Jax, K. (2003). Wofür braucht der Naturschutz die wissenschaftliche Ökologie? Die Kontroversen um den Hudson River als Testfall. *Natur und Landschaft*, 78, 93–99.

Jax, K. (2010). *Ecosystem functioning.* Cambridge University Press, Cambridge.

Jax, K. (2011). History of ecology. In: *Encyclopedia of life sciences (ELS).* John Wiley & Sons, Chichester, www.els.net/, https://doi.org/10.1002/9780470015902.a0003084.pub2

Jax, K., Barton, D.N., Chan, K.M.A. et al. (2013). Ecosystem services and ethics. *Ecological Economics*, 93, 260–268.

Jepson, P. and Whittaker, R.J. (2002). Histories of protected areas: internationalisation of conservationist values and their adoption in the Netherlands Indies (Indonesia). *Environment and History*, 8, 129–172.

Jones, M. (2003). The concept of cultural landscape: Discourse and narrative. In: *Landscape interfaces* (eds. Palang, H. and Fry, G.). Kluwer, Dordrecht, pp. 21–51.

Jones, K. (2012). Unpacking Yellowstone. The American national park in global perspective. In: *Civilizing nature. National parks in global historical perspective* (eds. Gissibl, B., Höhler, S. and Kupper, P.). Berghahn Books, New York, pp. 31–49.

Kareiva, P. (2014). New conservation: setting the record straight and finding common ground. *Conservation Biology*, 28, 634–636.

Kareiva, P. and Marvier, M. (2012). What is conservation science? *BioScience*, 62, 962–969.

Kinchy, A.J. (2006). On the borders of post-war ecology: struggles over the Ecological Society of America's Preservation Committee, 1917–1946. *Science as Culture*, 15, 23–44.

Kingsland, S.E. (1985). *Modeling nature. Episodes in the history of population ecology*. University of Chicago Press, Chicago.

Kirchhoff, T. (2019). Abandoning the concept of cultural ecosystem services, or against natural–scientific imperialism. *Bioscience*, 69, 220–227.

Knapp, G. (2017). Human Ecology. In: *International encyclopedia of geography* (eds. Richardson, D., Castree, N., Goodchild, M.M., Kobayashi, A., Liu, W. and Marston, R.A.). John Wiley & Sons, New York, pp. 1–9.

Knaut, A. (1990). Der Landschafts–und Naturschutzgedanke bei Ernst Rudorff. *Natur und Landschaft*, 65, 114–118.

Knaut, A. (1993). Zurück zur Natur! Die Wurzeln der Ökologiebewegung. *Jahrbuch Naturschutz und Landschaftspflege, Supplement*, 1, 1–480.

Kupper, P. (2009). Science and the national parks: a transatlantic perspective on the interwar years. *Environmental History*, 14, 58–81.

Kupper, P. (2014). *Creating wilderness: a transnational history of the Swiss National Park*. Berghahn Books, New York, Oxford.

Lee, E. (2016). Protected areas, country and value: the nature-culture tyranny of the IUCN's Protected Area Guidelines for indigenous Australians. *Antipode*, 48, 355–374.

Lekan, T. (2004). *Imagining the nation in nature: landscape preservation and German identity*. Harvard University Press, Berkeley.

Lenzing, A. (2003). Der Begriff des Naturdenkmales in Deutschland. *Gartenkunst*, 15, 4–27.

Leopold, A. (1939). A biotic view of land. *Journal of Forestry*, 37, 727–730.

Leopold, A. (1949). *A sand county almanac and sketches here and there*. Oxford University Press, New York.

Loo, T. (2006). *States of nature: conserving Canada's wildlife in the twentieth century*. University of British Columbia Press, Vancouver.

Mann, C.C. (2005). *1491. New revelations of the Americas before Columbus*. 2nd ed. Vintage Books, New York.

Marsh, G.P. (1864). *Man and nature*. University of Washington Press, Seattle, London.

Marvier, M. (2014). New conservation is true conservation. *Conservation Biology*, 28, 1–3.

Matagne, P. (1998). The politics of conservation in France in the 19th century. *Environment and History*, 4, 359–376.

McCormick, J. (1989). *The global environmental movement: reclaiming paradise*. Belhaven Press, London.

McIntosh, R.P. (1985). *The background of ecology. Concept and theory*. Cambridge University Press, Cambridge.

Meadows, D.H., Meadows, D.L., Randers, J. et al. (1972). *Limits to growth: a report for the Club of Rome's project on the predicament of mankind*. Potomac Associates, London.

Meine, C. (2013). Conservation movement, historical. In: *Encyclopedia of biodiversity* (ed. Scheiner, S.). Elsevier, Amsterdam, Vol. 2, pp. 278–288.

Meine, C. (2014). What's so new about the "new conservation"? In: *Keeping the wild. Against the domestication of the earth* (eds. Wuerthner, G., Crist, E. and Butler, T.). Island Press, Washington D.C., pp. 45–54.

Meine, C., Soule, M. and Noss, R.F. (2006). "A mission-driven discipline": the growth of conservation biology. *Conservation Biology*, 20, 631–651.

Millennium Ecosystem Assessment (2005). *Ecosystems and human well-being: synthesis*. Island Press, Washington D.C.
Muir, J. (2019/1901). *Our national parks*. Dover Publications, Mineola, New York.
Nash, R.F. (2014). *Wilderness and the American mind*. 5th ed. Yale University Press, New Haven.
Neumann, R.P. (2005). *Making political ecology*. Routledge, Oxon, New York.
Norton, B.G. (1986). Conservation and preservation: a conceptual rehabilitation. *Environmental Ethics*, 8, 195–220.
Norton, B.G. (1991). *Toward unity among environmentalists*. Oxford University Press, New York.
Noss, R., Nash, R., Paquet, P. et al. (2013). Humanity's domination of nature is part of the problem: a response to Kareiva and Marvier. *BioScience*, 63, 241–242.
O'Neill, J., Holland, A. and Light, A. (2008). *Environmental values*. Routledge, London.
Oelschlaeger, M. (1991). *The idea of wilderness*. Yale University Press, New Haven, London.
Ott, K., Potthast, T., Gorke, M. et al. (1999). Über die Anfänge des Naturschutzgedankens in Deutschland und den USA im 19. Jahrhundert. *Jahrbuch für Europäische Verwaltungsgeschichte*, 11, 1–55.
Pascual, U., Balvanera, P., Christie, M. et al. (eds.) (2022). *Summary for policymakers of the methodological assessment report on the diverse values and valuation of nature of the intergovernmental science-policy platform on biodiversity and ecosystem services*. IPBES Secretariat, Bonn.
Peterson, A.L. (2001). *Being human. Ethics, environment and our place in the world*. University of California Press, Berkeley, Los Angeles.
Pickett, S.T.A., Parker, V.T. and Fiedler, P.L. (1992). The new paradigm in ecology: implications for conservation biology above the species level. In: *Conservation biology. The theory and practice of conservation, preservation and management* (eds. Fiedler, P.L. and Jain, S.K.). Chapman & Hall, New York, pp. 65–88.
Pielke Jr., R. (2007). *The honest broker. Making sense of science in policy and politics*. Cambridge University Press, Cambridge.
Potthast, T. (2000). Bioethics and epistemic-moral hybrids: perspectives from the history of science. *Biomedical Ethics*, 5, 20–23.
Radkau, J. and Uekötter, F. (eds.) (2003). *Naturschutz und Nationalsozialismus*. Campus-Verlag, Frankfurt/Main.
Robbins, P. (2012). *Political ecology: a critical introduction*. Wiley & Sons, New York.
Roberts, J. (2020). Political ecology. In: *The Cambridge encyclopedia of anthropology (online)* (eds. Stein, F., Lazar, S., Candea, M. et al.). University of Cambridge, Cambridge. http://doi.org/10.29164/20polieco
Rolston, H.I. (1990). Biology and philosophy in Yellowstone. *Biology and Philosophy*, 5, 241–258.
Ross, C. (2015). Tropical nature as global *patrimoine*: imperialism and international nature protection in the early twentieth century. *Past & Present*, 226, 214–239.
Rudorff, E. (1880/1990). Über das Verhältnis des modernen Lebens zur Natur. *Natur und Landschaft*, 65, 119–125.
Rudorff, E. (1904). *Heimatschutz*. 3rd ed. Verlag von Georg Müller, München and Leipzig (first edition 1901).
Runte, A. (1997). *National parks. The American experience*. 3rd ed. University of Nebraska Press, Lincoln, Nebraska.

Sagoff, M. (2013). What does environmental protection protect? *Ethics, Policy and Environment*, 16, 239–257.

Sandbrook, C. (2015). What is conservation? *Oryx*, 49, 565–566.

Schlimm, A. (2020). Eine "Entente Cordiale" für den Schutz der Heimat? Ambivalente Europäisierung unter Landschafts–und Heimatschützern um die Wende zum 20. Jahrhundert. In: *Ambivalenzen der Europäisierung* (eds. Frysztacka, C., Timm, B., Weber, C. and Worschech, S.). Franz Steiner Verlag, Stuttgart, pp. 37–48.

Schmoll, F. (2004). *Erinnerung an die Natur. Die Geschichte des Naturschutzes im deutschen Kaiserreich*. Campus, Frankfurt, New York.

Schmoll, F. (2005). Indication and identification. On the history of bird protection in Germany, 1800–1918. In: *Germany's nature: cultural landscapes and environmental history* (eds. Lekan, T.M. and Zeller, T.). Rutgers University Press, New Brunswick, pp. 161–182.

Schwarz, A. (2011). History of concepts for ecology. In: *Ecology revisited: reflecting on concepts, advancing science* (eds. Schwarz, A. and Jax, K.). Springer, Dordrecht, pp. 19–28.

Sellars, R.W. (1997). *Preserving nature in the national parks: a history*. Yale University Press, New Haven.

Sheail, J. (1976). *Nature in trust: the history of nature conservation in Britain*. Blackie & Son, Glasgow, London.

Sheail, J. (2010). *Nature's spectacle. The world's first national parks and protected places*. Earthscan, London.

Soper, K. (1995). *What is nature?* Blackwell, Oxford.

Soulé, M. (1985). What is conservation biology? *BioScience*, 35, 727–734.

Soulé, M. (2013). The "new conservation". *Conservation Biology*, 27, 895–897.

Soulé, M. (2014). Also seeking common ground in conservation. *Conservation Biology*, 28, 637–638.

Spence, M.D. (1999). *Dispossessing the wilderness: Indian removal and the making of the national parks*. Oxford University Press, New York, Oxford.

Sutton, M.Q. and Anderson, E.N. (2013). *Introduction to cultural ecology*. AltaMira Press, Lanham.

Takacs, D. (1996). *The idea of biodiversity: philosophies of paradise*. John Hopkins University Press, Baltimore.

Talbot, L.M. (1983). IUCN in retrospect and prospect. *Environmental Conservation*, 10, 5–11.

Tallis, H., and Lubchenco, J. (2014). A call for inclusive conservation. *Nature*, 515, 27–28.

Teherani-Krönner, P. and Glaeser, B. (2020). Human-, Kultur- und Ethnoökologie. *Natur und Landschaft*, 95, 407–417.

Trepl, L. (1983). Ökologie–eine grüne Leitwissenschaft? Über Grenzen und Perspektiven einer modischen Disziplin. *Kursbuch*, 74, 6–27.

Tyrrell, I. (2012). America's national parks: the transnational creation of national space in the progressive era. *Journal of American Studies*, 46, 1–21.

van der Windt, H.J. (2012). Biologists bridging science and the conservation movement: the rise of nature conservation and nature management in the Netherlands, 1850–1950. *Environment and History*, 18, 209–236.

van Dyke, F. and Lamb, R.L. (2020). *Conservation biology: foundations, concepts, applications*. Springer, Cham.

von Droste, B., Plachter, H. and Rössler, M. (eds.) (1995). *Cultural landscapes of universal value*. Gustav Fischer, Jena, Stuttgart, New York.
Weiner, D.R. (1988). *Ecology, conservation, and cultural revolution in Soviet Russia*. Indiana University Press, Bloomington, Indianapolis.
Wiens, J.A. and Hobbs, R.J. (2015). Integrating conservation and restoration in a changing world. *BioScience*, 65, 302–312.
Williams, B. (1985). *Ethics and the limits of philosophy*. Fontana Press, London.
World Commision on Environment and Development (1987). *Our common future*. Oxford University Press, Oxford, New York.
Worster, D. (2008). *A passion for nature. The life of John Muir*. Oxford University Press, Oxford.
Wuerthner, G., Crist, E. and Butler, T. (eds.) (2014). *Keeping the wild. Against the domestication of the earth*. Island Press, Washington D.C.

3
ANALYSING CONSERVATION CONCEPTS

This chapter constitutes the core of the book. Any attempt to analyse or even "classify" conservation concepts becomes more complex and difficult the closer one looks. There are multiple ways to analyse conservation concepts and to arrive at some guidance through the maze of the many different, but often overlapping concepts in conservation and conservation biology. In Section 3.1, I will discuss various (non-exclusive) criteria and approaches for distinguishing different conservation concepts. These are especially the objects at which the concepts aim, the types of values related to specific concepts, the often implicit (and therefore neglected) strategic dimensions of conservation concepts, and finally the narratives of human–nature relationship to which the specific concepts link. As the starting point for further analysis, I will focus on the narrative dimensions of concepts, which I will develop in detail in Section 3.2. To reconstruct such implicit narratives, a general structure for narratives on human–nature relationships (in a conservation context) will be the developed. Based on this general structure, several "major narratives" of human–nature relationships in conservation will be carved out, as kind of "ideal types" (Section 3.3); they will be related to specific paradigmatic concepts, e.g. wilderness, cultural landscapes, biodiversity, ecosystem services. The final subchapter (Section 3.4) will open up the spectrum of concepts again, describing not only the major narratives and overarching conservation concepts but also other related concepts with which they form what I call "conceptual clusters". Further, I will highlight some important supporting, often overarching concepts (such as ecosystem functioning or the Anthropocene concept). Intaking up again the criteria discussed in Section 3.1, a number of tables will, taken together, provide a kind of conceptual matrix for distinguishing conservation concepts.

DOI: 10.4324/9781003251002-3

3.1 Different criteria and approaches for analysing conservation concepts

Analysing conservation concepts is a task that can easily lead into major confusion. Science and society are interwoven here in a complex manner. One could distinguish conservation concepts by means of several criteria such as the following:

- By the types of objects of conservation: *What is conserved?*
- By the types of values that are implied: *Why is conservation pursued? Why is an object conserved?*
- By the strategies implied that are used to justify conservation: *How to best communicate and achieve conservation (goals)?*
- By their use as tools to implement conservation goals: *How to best achieve conservation goals in practical terms?*
- By the main narratives of human relationships with nature in general and those behind conservation concepts in particular: *What is the story of the relationships between humans and nature?*

The same object (e.g. a particular species or an ecosystem) may be protected for various different purposes and on behalf of different underlying values, and it may – partly dependent on purposes and values – be protected by means of different strategies and tools. This means, the different categories listed above are neither clear-cut nor mutually exclusive. A particular concept may refer to several categories, e.g. "restoration" may denote both a tool and a strategy, or "biodiversity" both an object and a programmatic approach.

For conservation concepts (as well as for many ecological concepts) another difficulty is that even the *terms*, in their origin, come from different sources: some from science, some from politics, some from everyday life, often metaphorical (Larson 2011); and even within one and the same expression there may be a mix of scientific and other ones (e.g. in the term "ecosystem services"). Also, meanings can shift during the history of a concept, can be broadened or narrowed (Schwarz 2011; see Section 2.5). Meanings may "split" into different ones for one and the same term, maintaining only a common core but nevertheless telling different stories (e.g. the use of "ecosystem services"; see Section 3.4).

In this book, narratives of human–nature relationships will be used as the major explorative tool to structure the field of conservation concepts. Before turning to that, however, at first the other categories mentioned will be described, which, to a considerable degree, are integrated in such narratives, and to which I will come back in Section 3.4.

Objects of conservation

What does conservation protect? The short historical overview in Section 2.2 was roughly structured along these lines. There are even more objects of conservation. The most common are as follows:

- Individual features: (non-human) organisms, but also individual non-living objects such as mountains, rocks, or geothermal features. This overlaps with animal protection and with the protection of natural monuments.
- Populations of species
- Species itself (a difficult category for protection; see Box 3.5 in Section 3.3)
- Biodiversity (with different meanings and facets)
- Biotic communities
- Ecosystems
- Landscapes (including cultural landscapes)
- (Natural) processes
- Wilderness (if it is an object at all)
- Ecosystem services
- Natural resources

The matter is complicated by the fact that two different *types of objects* can be distinguished, related to the roles they play in conservation. I will call them the *pragmatic object* and the *ultimate object*. By the *pragmatic object* I mean the object that is usually, at first sight, seen as the object of conservation. It is that object at which conservation measures directly aim (e.g. stabilising a population or maintaining ecosystem services). The *ultimate object* of conservation is that object for whose sake, *for which* conservation is ultimately done (e.g. for the sake of species or for human well-being). The categories of objects may coincide in specific situations but they often do not. Mingling these categories adds to confusion about different conservation approaches.

An example: species conservation often protects specific species (e.g. the wolf, the peregrine falcon, or some rare plant species). These species are the pragmatic objects. However, the ultimate goal of protecting these species can often be either humans and their well-being (for a variety of reasons, e.g. to preserve potential medical resources) or the respective species themselves (e.g. if an intrinsic value – independent of human utility – is assumed). Likewise, an ecosystem can be the pragmatic object of conservation, but the ultimate object to be protected might either the ecosystem itself, specific species which depend on the system, or natural resources for human well-being.

One may argue that conservation wants to protect several of the objects mentioned together, and that, e.g., protecting ecosystems at the same time protects species (and their diversity), and vice versa. This is, however, not always the case. Trade-offs occur between species protection and the protection of ecosystems or ecosystem services (Jax and Heink 2015) or even between protecting different types of species (see case study at the end of this section).

For some of the objects listed above is not always clear how they are understood in specific settings. How to define and delimit "species", but also communities and ecosystems, is and has been the subject of much controversy (Hull 1997, Jax 2006, 2007). Likewise: What is wilderness? Is it an area, a process, a relation? What do we mean by biodiversity? Only "natural" diversity or also "artificial" diversity, including exotic species (see Angermeier 1994)? Many of these issues remain fuzzy and/or contested in discussions on conservation. Section 3.4 will discuss these questions for specific concepts.

Values

There are different justifications for protecting specific types of objects. These are often expressed via values. In conservation science, and even more in the practice of nature conservation, values are at the core of many controversies. Values are thus all over the place in conservation, and they are often implicit in conservation concepts. There is an extensive literature on values in conservation (e.g. O'Neill et al. 2008, Eser et al. 2014, Pascual et al. 2022). Recently, also the Intergovernmental Panel on Biodiversity and Ecosystem Services (IPBES) has been undertaking a "Values assessment" (see Pascual et al. 2017, 2022). Even the Convention on Biodiversity (CBD) starts its preamble with a long list of the various values of biodiversity, ranging from its "intrinsic value" via "ecological, genetic, social" values up to economic and recreational values.

Values can be described as "the conscious or unconscious standards of orientation and guiding ideas by which individuals and groups let themselves be guided in their choices of action" (Horn 1997, p. 332, translation KJ).[1]

Within conservation and environmental ethics, a distinction often has been made between an anthropocentric approach, embracing use values (instrumental values), and an ecocentric (or physiocentric) one, focusing on the "intrinsic values" of nature, values independent of any human interests. In such controversies, the ecosystem services concept is, e.g., considered as a strongly anthropocentric concept and the biodiversity concept as an ecocentric one.

Concerning intrinsic values, many discussions within environmental ethics deal with the question of which objects (beyond humans) have intrinsic value and for what reasons (e.g. Gardiner and Thompson 2017). Is it only sentient animals (because they can suffer)? Do all living beings have intrinsic value (on behalf of their interests to live and prosper)? Does it also relate to ecosystems (e.g. on behalf of their "integrity") or nature as a whole (simply because of its existence or its sacredness) (Gorke 2000)?

The juxtaposition of "anthropocentric" and "ecocentric" values as a dichotomy, however, is much too simple and does not do justice to conservation concepts such as biodiversity, ecosystem services, or wilderness. Nor does it really grasp the full spectrum of values of nature. Especially since the last decade, the discussion has broadened to include so-called "relational values", which form a kind of middle ground between the extremes mentioned (Figure 3.1). They denote those non-use values that people have with nature on behalf of the relation (instead of the object itself) towards nature and its components (Chan et al. 2016, Himes and Muraca 2018, and see Section 5.3). It is the relation itself that counts. This means that nature (e.g. some landscape) is valued for the role it plays in shaping peoples' identity, or affection to specific objects with which persons have a history. It also is expressed in attitudes of care towards individual organisms. These relations are non-consumptive, but they also do not depend on the notion of an "intrinsic value" of nature, independent of any human interest. Nevertheless, relational values are clearly "anthropocentric" as they relate to humans. In contrast to instrumental values, however, they are not substitutable as they aim at the specific. Relational values also draw on an old philosophical tradition, going back to Aristotle, that of eudaimonia. Values understood under an eudaimonistic perspective (as part of relational values) not only refer to a surplus in quality of life, in terms of leisure and aesthetic experiences, such as a walk in the woods or swimming in a natural pond. Rather and more properly, they refer to all those entities and processes

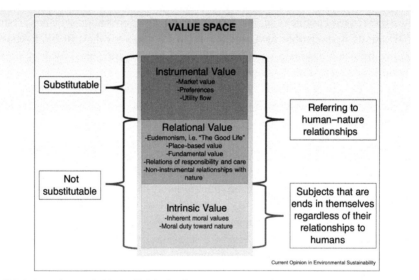

FIGURE 3.1 Distinction between instrumental, relational, and intrinsic values of nature.

Graphics reprinted from Himes, J. and Muraca, B. (2018). Relational values: the key to pluralistic valuation of ecosystem services. *Current Opinion in Environmental Sustainability*, 35, 1–7, by permission of Elsevier.

considered as necessary for living a "good life". It was such a notion of "good life" (truly fulfilled and human life) which was referred to by Greek philosophers as eudaimonia. From this point of view, eudaimonistic values are not confined to subjective preferences but extend to issues of intra- and intergenerational justice (Muraca, 2011). Moreover, beings holding eudaimonistic values are valuable in themselves and are not simply reducible to the benefits and services that they deliver as means. Thus the old oak tree in my garden, planted by my grandfather, is not an instrument to produce wood or sequester carbon – even if it can be used for that goal – instead it bears a deeper meaning for my life and cannot be simply replaced by something else. Figure 3.1 summarises these different types of values as currently discussed in conservation ethics and conservation biology.

A similar classification of values and ethical approaches towards nature was developed by Eser et al. (2014). Their terminology and classification is still a little closer to human behaviour and conservation practice. In an analysis of ethical reasoning in European national biodiversity strategies, Eser and colleagues distinguish three types of morally relevant arguments for conservation, which they labelled prudence, justice, and the good life. The "good life" largely corresponds to the relational, especially eudaimonistic values. "Justice" refers both to arguments of the intrinsic values – and potentially rights – of non-human nature (ecological justice) but also to intra- and intergenerational justice between humans. The latter addresses that some groups of humans can be affected in an unjust way by other humans' actions on nature (e.g. by clearing tropical forests at the detriment of indigenous people); it is often called environmental justice (see Section 5.2). Finally, "prudence" refers to all those arguments and values in favour of nature that are based on the prudential handling of nature (instrumental values), in a way that does not counteract general societal interests. One should not cut the branch on which one is sitting, as the literal translation of the German proverb says. This is the type of value which mostly is captured by ecosystem services approaches, but it also comes close to what is sometimes labelled "ecological values". There cannot be "ecological values" on the same categorial level as moral values, because ecology cannot provide values. What is described as "ecological values" in some of the literature is basically the observation that ecological insights (e.g. about species' requirements or ecosystem processes) can help to avoid imprudent actions on nature that would otherwise harm humans and/or other species. Prudential arguments and values are morally relevant as they do, ideally, not just refer to the interest of individual humans, much less the profits of companies, but to society as whole.

There are a couple of other value classifications which go into more detail (e.g. Kellert 1996), but I think that the rough distinction into three major types of nature-related values (instrumental, relational, and intrinsic values), as described above, is the most important one for understanding many controversies on values in conservation.

However, even though there is a strong focus on values in many conservation debates today, values are not enough for distinguishing major conservation

62 Analysing conservation concepts

concepts, because values are embedded: in actions, habits, history – and that is why, e.g., narratives come closer to everyday human experiences than (merely) abstract values.

Strategies

A number of conservation concepts, and those on which this book is focused on, are in fact designating whole conservation approaches, i.e., they have a programmatic character. One might also describe such approaches as "conservation strategies". But this is not what I mean here by "strategy" and "strategic".

Strategic aspects of conservation concepts are here understood in societal terms. They refer to the search for the best, most effective conservation concepts – but good and effective not meant in a technical, conservation-internal way. Strategic thinking of this kind reaches beyond the internal logic and value debates of conservation concepts but aims at the best way to communicate and bring forward conservation in a contested social setting. This can involve the use of particular arguments – or least a particular language – which conservationists might not embrace as their most favoured ones, but as that which appears to be most effective for reaching out to those one wants to convince of one's own goals. This can be, e.g., economic arguments, or, for John Muir, promoting tourism and recreation as an argument for wilderness protection, while in fact embracing nature's spiritual value (see Section 2.2). That is, the *main* function – albeit not always its only one – of such strategic arguments is to convince the interlocutor (or "opponent") of one's goals and not so much of one's own values. Strategic considerations, and the means to reach them, may be in line with the underlying values promoted by the acting persons or groups but may also be something that is mainly pursued in order to arrive at particular goals, even with means that do not completely coincide with one's underlying values.

Tools

Some conservation concepts are also – or only – tools for reaching conservation goals. Of the major concepts dealt below, one may consider, e.g., ecosystem management, "restoration", also as tools, also "protected areas". Concepts like population viability analysis or biological indicators are clearly analytical tools. I will not follow this aspect in much detail here, the more as concepts such as ecosystem management and restoration are much more than mere tools but are terms designating complex and in themselves highly heterogenous conservation approaches.

Case study: the Oostvaardersplassen experiment

An example where differences between ideas of conservation, namely about different target objects, different values, and different ideas of nature show very clearly is that of the Dutch conservation area called Oostvaardersplassen (OVP).

The Oostvaardersplassen, with an extension of 60 km², are part of large polder area, land reclaimed from the Ijsselmeer lake.[2] Reclamation was completed in 1968 and the area was originally designated for industrial development. This development was never realised and in consequence, the low-lying land was left on its own. With its remaining shallow water areas, wetlands, and spontaneous grasslands, it became very attractive for birds, not the least for migrating aquatic birds, among them some rare species, such as Eurasian spoonbills (*Platalea leucorodia*), but also rare songbirds like bluethroat (*Luscinia svecica*). Some parts of the area became designated for nature protection already in 1975, and in 1986 the whole area of what is now the OVP was given the status of a State Natural Monument (*Staatsnatuurmonument*). OVP became an experiment of the Dutch policy of "Nature Development", creating new nature, or what today would be called "rewilding". Under the guidance of ecologist Frans Vera the effort was to turn the new area into a wild place that should mirror how Europe was looking during the Pleistocene age, i.e. around 13,000 years ago. In what Vera (e.g. Vera 2009) described as a "paradigm shift", the idea was not, as the previous academic opinion had been, that Europe had been covered mostly by dense forest but, instead, that there had been a mixture of forested and open, non-forested areas, the latter resulting especially from the activities of large grazers such as the aurochs and the tarpan (the European wild horse) – both extinct today. To keep the land open in the imagined way, the closest functional equivalents to these grazers were introduced. Beginning in 1983, so-called Heck oxen (at first 34), rebreed from domestic cattle, and Konik horses (Figure 3.2), a small and robust race of domestic horses (at first 20) were released into the Oostvaardersplassen; 56 red deer followed in 1992 (van den Belt 2004). In contrasts to mainstream conservation in central Europe, there was no detailed management plan for the OVP, at least not until 2011 (Lorimer and Driessen 2014). The explicit philosophy for handling the area was one of letting nature take its course, i.e. of non-intervention. The domesticated animals should become feral over time (Vera 2009). Being surrounded by urban and agricultural areas, and even bordered by a highway and railway track, however, OVP is fenced, with no opportunity for the larger animals to move to other suitable areas. Not the least on behalf of this, the hands-off policy of the OVP management was contested early on and the Dutch government established an International Commission for the Management of the Oostvaardersplassen (ICMO), which delivered two reports, one in 2006, one in 2010.

Nevertheless, the area developed into a much-noticed example of a "new wilderness", with almost 80 birds species recorded, partly over 30,000 greylag geese (*Anser anser*) overwintering and grazing in OVP (Vera 2009) and a strongly developing herd of large grazers, reaching more than 5000 individuals at its peak. The former sea bottom is highly productive and allows for such a large number of grazers in the summer, but not always in winter, which resulted in starvation of animals during the cold season. Starvation was considered a natural process and taken into account by the OVP management early on. It caused, however, a huge

FIGURE 3.2 Konik horses in the Oostvaardersplassen area, the Netherlands. Photo: Ökologix, Public domain, via Wikimedia Commons.

outcry: among animal right activists, conservationists, and in the broad public. The non-intervention policy of the OVP management even led to parliament debates as well as to (unsuccessful) lawsuits against the responsible agency (the *Staatsbosbeheer*, State Forestry Service) (Lorimer and Driessen 2014). Mass starvation occurred in some harsh winters, with more than 900 dead animals in winter 2011/2012 alone. Since at least 2014, part of these deaths was due to shooting the weakest starving animals to prevent further suffering. In the winter 2017/2018 between 2000 and 3000 deer, Heck-oxen, and Konik horses died (Kopnina et al. 2019). After more public protests, this eventually led to the end of the non-intervention policy at OVP. From 2018 on, the government of the Province of Flevoland (where OVP are located) enforced a new management, jointly with the *Staatsbosbeheer*, setting target populations, at 210 Heck-oxen, 550 Konik horses, and 500 red deer (Gordon et al. 2021). Surplus animals are now shot or relocated to other places.

The controversy about the management of the Oostvaardersplassen has drawn much attention from different perspectives, on behalf of what can be learned from it for rewilding projects (e.g. Gordon et al. 2021, Kopnina et al. 2019), for the conduct of science and environmentalism and the relation between conservation and society (Lorimer and Driessen 2014), or from the perspective of environmental ethics (Klaver et al. 2002). I will focus here on the question of what role different target objects, different values, but also different ideas about nature have played in this controversy, specifically within the broader conservation community.

The most contentious issue was the fate of the large grazing animals. For animal rights activists they had intrinsic value, and avoiding suffering was the paramount duty, which in their eyes was violated. OVP managers, however, had another focus: for them the most valued object was the community or the ecosystem, in which large grazers performed, at least initially, a particular *functional* role (creating and maintaining the desired mosaic of habitats). They were introduced into the, at first, rather spontaneously developing area to form a "more *complete* naturally functioning ecosystem" (Vera 2009, p. 31, my emphasis), an ecosystem as it was imagined to have existed before humans were impacting the landscape. With such a focus, suffering of animals, as by starvation or diseases, is "normal" and part of the natural regulation of species. The conservation target is thus not the individual animals and even not the populations of particular species. Complaints were also uttered by, e.g., birders, demanding a focus on species protection, most urgently when, e.g., the rare spoonbill disappeared, a loss which, however, turned out as a transient (Lorimer and Driessen 2014; see also Lorimer 2015). Heck cattle and Konik horses were thus only means, the ultimate object being the ecosystem or "biodiversity", which was supposed to be highest in wild areas. The understanding of the ecosystem was guided by a systems perspective on ecosystems along the tradition of the Odum brothers (van den Belt 2004), where processes and functional roles mattered more than specific species or individuals. Thus, carcasses were considered as important for decomposition to take place as a natural process, also attracting birds like white-tailed eagles.

Much of the debate revolved around the idea of what is "natural" and what is "wild". Is only nature that is left undisturbed and unmanaged "real" nature, as OVP managers suggested, contrasting it to traditional Dutch conservation that focuses on cultural landscapes? Is real nature only nature without humans? But then, others argued, is the OVP really (close to) natural or wild, given not only the deliberate introduction of animals but also the fences around it? Can an area in a densely settled region be developed into something similar to wilderness? More than that, animal rights advocates argued that at least Heck cattle and Konik horses were not wild, having been introduced from zoos and farms. Do they really become "wild" after some time, some generations, do they become "de-domesticated", as Klaver et al. (2002) expressed it? A lawsuit against the OVP management decided that the OVP managers did not exert an "factual power" over the animals any more, that they ceased to be property and thus were *legally* wild (Lorimer and Driessen 2014). In consequence, OVP managers were not seen as responsible for, e.g., feeding the oxen and horses or for vaccinating them against diseases. Did that make the animals "de-domesticated" also in a factual and moral sense, i.e. that they were biologically not dependent on human care anymore and thus not subject to the moral obligations given for domestic animals? Klaver and colleagues, in their thoughtful paper from

2002, have tried to answer this question. They came to the conclusion that the OVP and the large grazers therein form a kind hybrid between wild and domesticated, natural and "artificial". Seeing "natural" as a gradient rather than one side of a dichotomy, they emphasise that de-domestication is a long learning process and that within that process, the responsibility to the individual may be replaced by what they call respect for *potential* wildness. The learning process here meant a learning process both for the animals as well as for the management – learning how to support the very process of de-domestication, not avoiding any intervention but also not treating the ungulates like domestic animals.

In response to the critique of the large mortality of the grazers, Vera (2009) rightly pointed out such high mortality has also been observed in "real wild" settings as, e.g., the Serengeti. The argument that this mortality is "unnatural" and mainly due to presence of fences and a lack of top predators (e.g. wolves) is not supported by studies "in the wild". The famous long-term investigation of the dynamics of moose and wolves on Isle Royale (Peterson 1995, 1999) has shown that major population crashes of ungulates due to disease and starvation also occur in much larger areas, with Isle Royale having an extent of 535 km^2, almost nine times that of the Oostvaardersplassen.

While the OVP experiment was described by some authors as "failed" (e.g. Theunissen 2019, Kopnina et al. 2019), other, such as Lorimer and Driessen (2014) or Schwartz (2019), judge more carefully. In many respects, it is a matter of what we value, how we, e.g., deal with the tension between the ideas of the "wild" and the well-being of individual animals, especially in a setting highly exposed to the public. Is a "wild" nature, with animals suffering and carcasses remaining, the kind of "naturalness" that conservationists and other people want? It at least disturbs the idealised public image of nature as mainly benign and harmonious. Much thus depends on our general idea of nature: how much management is either *permissible* to not spoil "real" nature or – on the opposite side of the spectrum – how much management is *necessary* to maintain particular valued target objects, such as specific species or a high number of species (see also Sarkar 1999).

3.2 Narratives as a structuring tool

> In the beginning was the story. Or rather: many stories, of many places, in many voices, pointing towards many ends.
>
> *(Cronon 1992, p. 1347)*

> Scientific theories […] begin as imaginative constructions. They begin, if you like, as stories, and the purpose of the critical or rectifying episode in scientific reasoning is precisely to find out whether or not these are stories about real life.
>
> *(Medawar 1982, p. 53)*

Narratives and narrative analyses have been used in various ways in environmental and conservation contexts. They have been applied to understand historical and recent events perceived as environmental crises, such as the "dust bowl", and possible management consequences deriving from them (e.g. Cronon 1992, Robertson et al. 2000). Some authors also emphasise the potential of narratives for environmental ethics (e.g. Treanor 2008, Ryan 2012, O'Neill et al. 2008). From a more policy-oriented perspective, researchers use narrative analysis as a means to describe and understand conservation problems, e.g. with a perspective on complex networks of humans, their institutions, and non-humans (Lejano et al. 2013, Ingram et al. 2015). More recently, Louder and Wyborn (2020) have been arguing for the power of a more complex narrative analysis in conservation, in that it "can help with reflection upon and questioning of underlying epistemologies and ontologies [of conservation], opening up space for diverse understanding" (p. 257) of conservation. They also present a list of selected narratives and counter-narratives they see as important in today's conservation.

I decided to extract, or reconstruct, the – often implicit – narratives connected with or even underlying conservation concepts as the major starting point for my analysis. If, as said earlier, the ultimate goal of conservation is to strive for good human relations with nature, narrative accounts of human–nature relationships appear to be especially well-suited to grasp the differences between conservation concepts and strategies and to situate them in larger societal contexts.

While narratives have become quite fashionable as an analytic category in many research fields (especially the humanities), they possess different advantages which make them highly useful also for the purposes of this analysis. First of all, they connect many of the other categories listed above: they include specific objects of conservation, values, goals, but also scientific findings, and not the least personal experiences; they often also form a kind of justification for specific positions and modes of action and thus serve strategic purposes. Further, they are close to our everyday understanding of the world, are less abstract than mere scientific or philosophical categories. So I will use narratives as a first attempt to sketch, to reconstruct different types of human relationships with nature that (often implicitly) underlie conservation and see to which other categories they are connected.

The narratives that will be discussed are closely related to what people understand by "nature". This is not so surprising as "nature", as said earlier (Section 2.3), is not a scientific concept but a philosophical one, and it is a highly relational concept, being defined in relation as well as in contrast to us, as individuals or as society as a whole.

One can find explicit short narratives that illustrate and legitimise at least part of humans' relation with nature in books and papers of conservation. A simple example is Ehrlich's and Ehrlich's (fictional) story of the "rivet popper", by which they start their influential book *Extinction* (Ehrlich and Ehrlich 1981, p. xi ff), comparing the ongoing human-caused extinction of species to a person who pops out more and more rivets (symbolising the species) from an airplane wing, endangering the performance of the whole airplane (symbolising the biosphere), of which he is also a passenger. Another fictional example is the story told by Gretchen Daily (1997) by which she illustrates which "services" nature provides to us, using the example of a space station on the moon and what species it would take for humans to survive there, independently from further external inputs.[3] Nonfictional conservation narratives can be found, e.g., in the work of Aldo Leopold, especially the Sand County Almanac (Leopold 1949). His famous story "Thinking like a mountain", describes a major change of his personal relationship with nature during an encounter with a wolf that he had been hunting, and which he killed.

Quite often, narratives are seen as a communication instrument for conservation concerns, and "new" and "more effective" narratives are sought to convince the broader public about the need for protecting nature. This is not my intention here. I am instead using narratives as an analytical method to understand and order the human–nature relationships implicit in major conservation concepts, building on methods from the literature sciences.

In what follows, I will first explain in more detail what I mean by narratives, describe the main elements that constitute narratives, and explain how narratives may be used as an analytical tool in the current context.

What are narratives and how to analyse them?

There have been various attempts to define "narrative". The concept, almost obviously, originally comes from literature analysis, but in recent decades, at least since the 1980s (see Elliott 2005, p. 5), has also been applied to other fields, such as history, philosophy, or the political sciences – to name just a few (see Meuter 2014 for an overview of various disciplines where narration is an issue). Although definitions of narrative differ, they have in common that a narrative describes *a sequence of events and binds them into a coherent and meaningful whole* (Ryan 2007, Liszka 2003, Journet 1991).

Journet (1991, p. 448) writes:

> Narrative is the selection of "significant" events and the arrangement of those events into a "significant" sequence, the principle of "significance" being the writer's way of imposing meaning on experience.

For a closer look at narratives and for their analysis, it is useful to distinguish at least two levels, which I will call here, following Bal (2017), "fabula" and "story" (see Box 3.1 for other uses of major terms in narratology). A *fabula* is "a series of logically and chronologically related events that are caused or experienced by actors" (Bal 2017, p. 5), while the *story* is the way in which these events are recounted, arranged, filled with detail and direction. Although I here deal with stories mostly as expressed in written or oral form, they may also come as film, image, piece of music, or in yet other forms. We will see later, that this distinction between fabula and story is an important one. This is, inter alia, because it allows to separate what I see as the general structure of narratives about human–nature relationships from the specific stories by which they are expressed; these stories convey quite different meanings and moral messages to the events of the fabula. While the fabula provides events in a simple chronological order, a story building on it will often shift with levels of time, e.g. start in the present, then go back to the past to explain the current situation, as is often the case in narratives of human–nature relationships.

Within a fabula, one can distinguish *events*, the things that happen, and *actors*, or broader: *actants*. The latter term, which I will preferably use when dealing with a fabula, does not so much allude and is not restricted to specific persons or characters (which occur in the story) but can also be filled by non-human "actors". These can be groups of people, even networks of people and institutions (Lejano et al. 2013), but also – in the case of conservation – animals, plants, or even nature as a whole. Actants can also be mere "roles" in the structure of a fabula, a functional role that a class of actors performs (Bal 2017, p. 166f). These actants are then specified and characterised in a specific story which relates to the respective fabula.

A narrative must thus have a fabula (a sequence of causally linked events) with a beginning and – ideally – an ending (closure), and it must transport some message expressed in and derived from the story. It must "communicate something meaningful to the audience" (Ryan 2007, p. 29). Often this message has also a normative content, expresses a lesson learned and something desirable; it thus involves an outlook to the future. Liszka (2003) writes:

> That which ultimately makes something into a story – understood as a series of events with an ending – is simultaneously what makes it normative. […] There is a certain teleological logic at work here, such that at the thematic level we move beyond explanation of events to *what such events might mean, normatively speaking.*
>
> *(p. 48)*

BOX 3.1 TERMINOLOGY AND METHODOLOGY OF NARRATOLOGY

There is no standard terminology in narratology, i.e. the approach of narrative inquiry. Even more, the same terms are used for different things, so confusion easily arises. Thus, what I call, following Bal, "fabula" is often called "story", while what I call "story" is often termed "plot" or "narrative discourse" (e.g. Abbott 2008). Likewise another term for "actant" may be simply "entity" (Abbott 2008). For overviews on the whole field of narratology, see the textbooks by Bal (2017) and Abbott (2008), for an effort to at least partly sort out the terminology, see Franzosi (1998). Another in-depth source is the *Handbook of Narratology* (Hühn et al. 2014).

Discussing the narrative approach to history, White (1980) even postulates that it is only by moral messages that some kind of closure to a sequence of real events is reached, as real events, i.e. history, in our common understanding never find a real ending but will continue with new, other events.

The message from the rivet popper story of the Ehrlichs', e.g., is simple and clear. If humanity continues to drive more and more species to extinction, the Earths' life support system (the biosphere or Earth's ecosystems, here the airplane) will eventually reach a tipping point and collapse – at the detriment of both humans and non-human beings. The obvious normative message is that humanity must change its behaviour to avoid such a catastrophe, in its own interest.

Human–nature relationships: an overarching structure

Human–nature relationships are fundamental, not just for characterising how we think about and act on nature, but also as something that is an important part of defining our own identities as humans. Human identity is often highly narrative (Ricoeur 1992, Bamberg 2014). Just think of how people commonly introduce each other: by telling where they come from, where they moved to, where and when they were educated, which profession they chose, etc.

So if we think of human–nature relationships in terms of narratives, such narratives integrate their *explanation* (how did this relationship arise?), a *justification* (why should we act?) and an *evaluation* (how good is the status quo of this relationship?) connected with a *goal* (what is a desired, proper relationship with nature?). Elaborating human–nature relationship(s) is thus not just describing how it (they) came about but much about giving *meaning* to our past and current interactions with nature.

Human relationships with nature can be best expressed with a couple of distinct stages of the fabula. These create an overarching structure that should allow to fit in various and content-wise quite divergent narratives (stories) on human–nature relationships.

The scheme I present here was derived largely deductively together with my own experiences of work in ecology and conservation biology. The aim of the scheme it to provide a systematic, comparative approach towards narratives of human–nature relationships. It should be emphasised that the scheme is primarily meant to characterise human–nature relationships as *expressed by those dealing with conservation and conservation biology*, and not necessarily as characterising *all* possible human–nature relationships of people. Also, the scheme is a heuristic device and not a sociological analysis. As presented here, it is a suggestion, open for modification and further development.

I suggest to characterise the general structure of human–nature relationships by a fabula consisting of the following seven stages:

1 The world ("nature" and "humans") coming into being
2 Initial state: humans living in/from/with nature
3 Development/change of human interaction with nature

4 New state: humans living in/from/with nature
5 Crisis/disruption perceived
   [Stages 2–4 (or 2–5) can be repeated in iterations]
6 Desired/final state: desired roles of humans and reference state
7 [Anticipated course of how to arrive at this state]

The above list is what in the following will be called the overarching "masterfabula"[4] (or "skeleton") of human–nature relationships, understood as narrative. Very different ideas of human–nature relationships can be fleshed out via this structure, either still rather general types of relationships (the "major narratives" of Section 3.3) or specific stories.

Before I describe and discuss the different stages of the masterfabula, some further cautionary remarks are necessary. First of all, all stories, and in part even fabulas, are constructed. They do not simply depict "reality" (although the fabula is the level closest to it) but are an outcome of analysis of various stories and texts (text in the broadest sense, including art, music, movies, etc.). The relevant events and actants are always a preselection from the whole of our reality, here selected for the specific purposes of portraying human–nature relationships as narratives. Even the stages of a fabula could have been selected differently, but that then is a matter of debate: how to best structure what makes up such relationships.

Second, these narratives are complex and they are almost never expressed explicitly in the way described above; not every part is always realised in a specific story. On behalf of this, my analysis of conservation narratives (as elaborated in Section 3.3) is in fact more a *reconstruction* of ideal narratives.

Third, the threshold between reality and myth (or fiction) is not always easy to draw, especially when looking back into the past. But to draw this line very clearly may also not be so necessary. The "mythical" parts (e.g. the creation of the earth by God, or myths of growth and progress in modern societies) are – for those who adhere to it – as real as any scientific findings and they may be essential for a person's relationship to himself and to nature.

The elements and stages of the masterfabula for human–nature relationships

There are some basic elements of the masterfabula which recur in different stages. They are especially actants, events, and, in part, evaluation. What does this entail in the current context?

Actants. All narratives of human–nature relationships have at least *two main actants*: humans and nature. For analytical and semantic purposes, such a distinction is unavoidable at first, even if humans are seen as part of nature (see Sections 4.1 and 4.3). These main actants may be split up into smaller groups and change in their characterisation in various ways during the course of the fabula or in different stories of the same fabula. As already briefly discussed in Section 2.6, there is, even though we easily speak about "humans" and "nature", no universal humanity, at least in later stages of the fabula. Sometimes then, a distinction is between "traditional" or "indigenous" humans on the one hand and "modern" humans on the other. In other stories, this distinction will be even more fine-grained, e.g., between industrialists and conservationists, powerless and powerful groups, or the like. Likewise, often the other main actant may not be "nature" as a whole but "biodiversity" or individual species, or abiotic and biotic nature. This, as will be seen later, makes a decisive difference between different narratives of human–nature relationships. A third, optional, but often very important actant, in religious contexts, is God or various spiritual beings.

The different actants must be *characterised*, in a specific narrative (story), i.e. their properties have to be described. This partly happens via direct description, partly via their actions (see Abbott 2008, chapter 10, Jannidis 2014). Human actants thus turn into characters in a specific type of human–nature narrative but often remain on a supra-individual level as "humans", "indigenous people", "Europeans" etc., i.e. as types. I am using "characterisation" here also when it comes to describing the properties of non-human entities, such as "nature" or "biodiversity". Properties of nature are, even in the conservation literature, sometimes described in a highly evaluative – and even anthropomorphic – manner, such as when nature is characterised as "good", "hostile", "harmed", or "healthy" – or even when ecosystems are seen as "providers" of services.

Events, the second major element of narratives, are what happens. Bal (2017) defines events as "the transition from one state to another" (p. 5) and an *act* as "to cause or to experience an event." An act – under this definition – is thus not restricted to human actants. Also "nature" or parts of it that can "experience" an event (e.g. human impact) or cause one (e.g. an earthquake).

Evaluation plays an important role in any narrative (story), as it is indispensable in providing meaning to narratives. As Elliott (2005) emphasises, the evaluation is not just imposed by a narrator, but it is also, at least in part, the result of an evaluation that results from an interaction between narrator and reader/listener:

> It is the evaluation that conveys to an audience how they are to understand the *meaning* of the events that constitute the narrative, and simultaneously indicate what type of response is required. The evaluation should not therefore be understood as simply provided by the narrator; rather the achievement of agreement on the evaluation of a narrative is the product of a process of negotiation. While the speaker can be understood as responsible for producing a narrative with an acceptable evaluation, the addressee or audience must collaborate by demonstrating that the evaluation has been understood.
>
> *(p. 9)*

Specific expressions of the elements will become visible in the following detailed description of the seven stages of the scheme. This description should also show already how stories emerge from the masterfabula.

(1) The world ("nature" and "humans") coming into being

The narrative starts with an *event*, which brings the major actants (humans and nature) into being. The world comes into being either by itself or by an act of God (or gods). Going back in time that far may appear somewhat strange and of minor importance for describing human–nature relationships. But in fact this is not the case. Even though this stage is only implicit in many narratives of human–nature relationships, it can be of decisive importance. Especially in religious contexts, the creation of nature by God often involves a kind of responsibility and accountability of humans to protect nature – not by direct responsibility towards nature itself (as in ecocentrism) but indirectly by responsibility towards God/the creator (theocentrism); this is often implied in *stewardship* concepts (see Box 3.6). However, non-religious descriptions of how the natural world came into being, especially as a product of natural evolution, can also imply ideas of responsibility towards nature, e.g. in terms of *kinship* of humans with other living beings (e.g. Rozzi 1999). In a Western context, where "modern" conservation arose, the religious context is mostly one of Christianity – however also here with various different (and sometimes completely opposite) interpretations of the two creation stories in the book *Genesis* (see Box 4.2). I will come back to other narratives outside the mainstream of the Western conservation tradition in Section 4.3.

(2) Initial state: humans living in/from/with nature

To properly describe this stage, it is necessary to introduce some special features of narratives on human–nature relationships, characterising the actants and their interactions, which also will recur in the other stages. They will serve as important analytical tools throughout this book. Some more space will be devoted here to explicate them. These features are in particular how nature is perceived by the actants: *what/how is nature?*, the *overall type of relation/interaction with nature*, and the *roles* that humans have in that respect.

Type of relation/interaction with nature: The primeval state of human interactions with (non-human) nature, which stage 2 describes, can be characterised overall roughly by three options.
  This relation is seen as either:

- An ideal world, where there was no distinction or at least no antagonism between humans and nature
- A period of struggle and fight, where humans lived in an adversary nature with hardships to their lives and even survival, or – less common – as
- Nature struggling with the influence of humans

In all cases, the initial state of humans in nature is usually considered a state in which the relation persisted for a long time, as a kind of equilibrium or at least steady state.

74 Analysing conservation concepts

Roles of humans: Each of these three types of relation can be characterised in particular stories by various roles of humans, (as one of the main actants in that relation) in or deriving from, respectively, these relations. Such roles, which can be found in different stories about human–nature relationship, and implicitly in different conservation concepts (Section 3.3), are, e.g., co-inhabitant, steward, guardian, prudent user, victim, dominator, and even nuisance.

What/how is nature: The other main actant, nature, is characterised as well, on the one hand descriptively, on the other from the point of a (moral) evaluation. In conservation concepts and its related narratives of human–nature relationships, the descriptive characterisation of nature is the main place where science, where especially ecological knowledge, comes in. In a descriptive manner, one may ask as to whether nature is static or dynamic, resilient/robust or fragile/endangered, predictable, unpredictable, vulnerable, autonomous, etc. This is also contested within ecology from a theoretical point of view (e.g. balance of nature or flux of nature: Pickett et al. 1992) and may differ for specific parts of nature and for specific locations.

Already at this stage, the characterisation is not without *evaluation*. In actual narratives/stories, evaluation is often implicit. In the context of the masterfabula it is part of the *analysis* of narratives and occurs even if it is not made explicit.

In terms of characterising nature in an evaluative way, nature may be perceived as neutral, benevolent, nourishing, adverse, threatening, frightening, independent, awesome, revengeful, adorable, sacred, etc. These characterisations are not mutually exclusive. They also mirror some value typologies (e.g. Kellert 1996).

This characterisation of nature already has a decisive influence on how stage 3 (the further development of human interactions with nature) is seen: as degradation/decline of human–nature relationships, as progress, as emancipation from nature, etc. (see Box 3.2).

Like stage 1, stage 2 is often characterised partly by mythical ideas, at least implicitly (e.g. by the image of paradise).

In the initial, "primeval" stage, "humanity" is mostly seen as a unit, i.e. as not (or at least not much) differentiated into different social groups or "civilisations"

BOX 3.2 PROGRESS AND DECLINE

The idea of progress is deeply ingrained in today's Western thinking, with respect to society, human morality, as well as to human–nature relationships, to the point that it seems almost self-evident to many people. The idea is, however, not a kind of anthropological constant in human thinking, as already a glimpse into other cultures, e.g. Buddhism or Australian Aboriginal culture, shows. But even in Western cultures, the idea of progress is in fact a rather recent theme, gaining prominence only in the 18th century in the course of European Enlightenment and later in the beginning of industrialisation

(Rapp 1992, Lange 2019). Although being contested from the beginning, the idea of progress almost gained the character of a myth, promising to provide meaning and direction for human history.

Conservation scientists, often educated as biologists, sometimes have a scholarly affinity towards the general idea of progress, due to particular interpretations of evolution (see Larson 2011). But even in a biology, progress is a difficult and contested concept (Gould 1989, de Cesare 2019).

There are a couple of questions related to the notion of progress. An important distinction to be made is between a *descriptive* and a *normative* meaning of progress (Rapp 1992, p. 26f; Rapp uses the term "genetic" in a philosophical sense, for what I call "descriptive" here). "Progress" – as well as its antonym "decline" – describes relations, in terms of a comparison of states at different instances in time. In a descriptive meaning it simply refers to a progression of events, the development of something new, without attributing any value to this process. The normative meaning implies that every such progress is also a development to something better, i.e. is valued as positively. While both meanings are often used together, there is no necessary link between a descriptive and normative meaning, and it should be taken care not to equate them. "Decline" denotes also a kind of change, but in contrast to "progress" it is always a thick concept by which change is valued as negative. The same phenomenon/series of historical events, e.g. the French Revolution or draining swamps, might be described as either progress or decline, depending on which valuation scheme is applied.

Also, "progress" is often used as an all-encompassing, universal concept. The question, especially in terms of a normative meaning, must, however, be progress of *what*, assessed by *which criteria?* (Lange 2019). This is at the core of many discussions on progress today: does a progress in technology also imply a progress in human well-being, human rights, and/or in the morality of humans, including good (and sustainable) relations with nature? It is obvious from history and even from our everyday observations as political beings that these different goals or areas of progress often do not match and even may be in conflict.

In conservation, progress (not the least when associated with economic growth, but also technical progress; see. Schmelzer 2015) has often been valued negatively, as a cause of environmental destruction. But the idea of progress is far from "dead" (Seefried 2015) even after awareness of the "limits of growth" (Meadows et al. 1972) in the 1970s. Even hard-core conservationists wish or see some (long-term) progress, e.g., in the perceived extension of the circle of moral objects (from humans to animals, plants, and ecosystems) or simply social progress as increase in conservation efforts.

(but see Head 2014). The actant "nature" may, however, already at this stage be further subdivided into specific parts, such as inanimate and animate nature, animals ("good" and "bad" ones), plants, or groups of the same.

These first two stages are mostly used in looking back, for setting a reference state (e.g. for a still proper relation to nature).

(3) Development/change of human interaction with nature

In the next stage of the human–nature relationship fabula, the initial "equilibrium" or steady state is disrupted or at least changed; the change thus may also be positive. These changes may come about both by human actions but also through nature's "actions", e.g. changes in climate (such as by ice ages), earthquakes, or other developments.

The process may be *characterised* (rather neutrally) as shifting, but also by thick concepts (see Section 2.6) such as progress, emancipation, or decline/degradation. Of course the question must always be: progress/decline of what? emancipation from what/in which respect? (Box 3.2)

The latter terms are already carrying implicit evaluations of various kinds, and even the same term can carry different evaluations. While progress is often (if not mostly) valued positive, certain types of progress have, especially in a conservation context, been valued negatively, as leading away from a previously better state of nature and/or human–nature relationships. The evaluation can be made on the changes themselves but can also be closely linked to the outcomes of these changes (stage 4).

Within this development of human interaction with nature, also the roles of humans and the perspective on nature may change as specified in stage 4.

The changes in the interactions between humans and nature might be described – in contrast to the primeval state – already as leading to the division, differentiation, or fragmentation of the actant "humanity" into more or less distinct "civilisations", societies, or social groups, the apex of which is the coarse and often problematic description of human history as leading to a distinction between "modern" and "traditional" societies of humans ("man").

(4) New state: humans living in/from/with nature

In its structure for analysis, this stage of the narrative is identical with stage 2. What has changed, but in a significant way, are the details. The characterisations, as described for stage 2, and the corresponding evaluations of the same also are made again in this stage. Also, as said before, the actants might have changed in terms of a greater differentiation. Each new state of humans living in/from/with nature, not only the "primeval" one, might be a reference state for conservation in terms of an assumed good human–nature relationships in some stories.

It should be noted that stages 3 and 4 may occur several times in some narratives, as a kind of *iteration*: from the primeval state to an early state of gatherers and hunters, through a state of farmers, to one of urbanisation and industrial agriculture (see example fabula below). However at some point in the narrative, the perspective on the changes and the resulting new states leads to the perception of a situation (or an ongoing development) as a *crisis* of the human–nature relationship.

(5) Crisis/disruption perceived

The perception of a crisis in the interactions between humans and nature is at the origin of the conservation movement and of conservation biology. The diagnosis of such a crisis is the *raison d'etre* for conservation and conservation biology. Michael Soulé (1985, p. 727), e.g., wrote: "Conservation biology differs from most other biological sciences in one important way: it is often a crisis discipline".

Crisis demands action to overcome it. The same situation, the same state of nature and of human–nature interaction may, however, not be considered a crisis by some people, e.g., because they feel that conquering and using, or even exploiting

---

BOX 3.3 CRISIS

"Crisis" is a multifaceted concept, to say the least. The term is omnipresent in current political and social debates and for some authors it is even a buzzword without the necessary precision to be useful in an analytical context. Nevertheless, for the issue of nature conservation and human–nature relationships it still appears as an important and useful concept. The "ecological crisis", the "environmental crisis", the "biodiversity crisis", the "extinction crisis", and similar expressions are all over the place in publications about conservation. As quoted earlier, Michael Soulé in his seminal paper from 1985 also referred to conservation biology as a "crisis discipline". Soulé also explicitly compared conservation biology to cancer biology. This comparison between conservation and medicine, which is not uncommon in nature conservation and conservation biology, in fact meets one of the original meanings of "crisis". In ancient Greece "crisis", as a *medical term* related to illness, referred "both to the observable condition and to the judgment (*judicium*) about the course of the illness. At such a time, it will be determined whether the patient will live or die" (Koselleck 2006, p. 360). Thus, a crisis is not something that is necessarily bad or has a bad outcome but can initiate a healing catharsis as well.

Another important meaning, partly related to the former, is crisis as *moment of decision*, requiring urgent action: "'decision' in the sense of reaching a crucial point that would tip the scale" (Koselleck 2006, p. 358); this meaning was present likewise already in ancient Greek. In recent times, the use of crisis in this sense often comes "to conjure up an imminent threat, one that demands an immediate and drastic response" (Graf and Jarausch 2017). This sounds familiar in a conservation context.

> Instead of a particular moment in time, crisis can also denote the *process* leading to it as "a singular, accelerating process in which many conflicts, bursting the system apart, accumulate so as to bring about a new situation after the crisis has passed" (Koselleck 2002, p. 240). This means that crises often are seen as threatening an existing structure, system, or value(s) and are frequently related to an experience or fear of loss. Another meaning not uncommon in writings on conservation and the environment is a view of *crisis as leading to a final point in history*, a meaning that represents a (mostly) secularised version of apocalyptic thinking (Koselleck 2002, 2006; see Veldmann 2012 for the treatment of apocalyptical reasoning in environmentalism).
>
> These and other meanings of "crisis" are not always strictly separated, which contributes to the vagueness of the concept. Although I did not find any place in the conservation literature where "crisis" has been explicitly defined, the descriptions above, in my view, seem to capture quite well the various implicit meanings of "crisis" in the conservation literature.
>
> In epistemological terms, most authors today emphasise a close relation of crises to human perception, meaning that "crises do not exist in the world until they have been conceptualized as such by contemporaries or historical observers" (Graf and Jarausch 2017). This is a very important point, as crises (and the timing of their beginning) may only be described and labelled as such much later and not already by those who directly experience the respective situations and events. In this context, it is important to analyse by whom and for which purposes, with which intention, something is described as crisis.

nature is good or even a moral duty (e.g. for the sake of material human well-being or as a religious obligation).

The *timing of the crisis*, i.e. at what stage in history a major crisis *occurs*, is of major importance as it has consequences for a projected *reference state* of (good) human–nature relationships and the desired state of nature (see Box 3.4 and the discussion about when the "Anthropocene" might have begun). The time when the crisis is seen as having started is, however, not always identical with the point in time when it first has been *perceived* as such. Often, the beginning of the crisis is described as having taken place much earlier, and it is not described by those that directly experience the respective situation and event. A particular loss (of species, landscape structure, ecosystem services, etc.) might be perceived as problematic only long after it occurred.

Like stage 3 and 4, also stage 5 may be repeated if a particular human–nature relationship, and the state of nature related to it, is perceived as a crisis and prompts new developments (stage 3). Such developments can lead to modifying or even resolving the existent crisis or to an additional new crisis.

Stages 1 to 5 constitute the basic narratives of human–nature relationships in a conservation context. Stages 6 and – sometimes – stage 7 look into the future and describe, in a kind of conclusion, what a proper, morally desired human–nature relationship should look like.

(6) Anticipated/final state (future)

Stories of human–nature relationships are not just fictional stories, they are in essence historical narratives, which, as Fulda (2014, p. 228) states, "integrate[s] past, present and future". They use the past to understand the future. This future may be formed along the description of a (real or imagined) past or present, or it may be something completely novel.

Many narratives on human–nature relationships in fact present alternative "scenarios" of the future: a negative, even apocalyptical future (mostly as results of continuing "business as usual"), and a positive future, to which a reform of the relationships of humans with nature could lead.

In this stage, the narrative needs to characterise on the one hand the desired role(s) of humans with regard to nature as well as some kind of reference state. The latter may be taken from pre-crisis stages or be a completely new one.

Concerning the potential role of humans with respect to nature, new role types may be added, like that of the penitent or the healer (e.g. in restoration efforts; see below). A less prominent but interesting idea is that of the late philosopher Klaus Michael Meyer-Abich (1984) who saw the role of humans not the least as giving a voice to nature, by this allowing nature to come to itself.

(7) [Anticipated course of how to arrive at it]

This phase occurs only in some cases, because it is already part of specific political and strategical plans. Common, at least, are postulates for a new mindset, a shift in attitudes towards nature, often, however, without details of how this could happen.

Explicating the different stages of the masterfabula allows us to ask new and productive questions to different conservation concepts and approaches, not the least that of the (desired) roles of humans in relation to nature. In what follows, I will apply the scheme developed above with a short example, trying to set up a full fabula spanning all seven stages.

A (semi-)fictitious fabula of human–nature relationships in conservation

To better illustrate the different stages, I am presenting in the following a potential – yet plausible – narrative. Of course, others are possible. This narrative is still on the level of the fabula, i.e. the succession of events.

A fabula to start with

Stage 1: In the beginning, the world came into being by the Big Bang. Gradually, the development of the universe proceeded, and some 4 billion years ago,[5] life originated on planet earth. Biological evolution brought about the first primates some 50–100 million years ago and the first hominids some 20 million years ago. The appearance of current humans (*Homo sapiens*) is said to happen around 200,000–300,000 years before present, mainly in Africa.

Stage 2: Humans gradually spread to almost all continents (Antarctica excluded). Humans were dependent on nature as any other species.

Stages 3 and 4 (in iterations): Humans developed tools to modify the environment for their needs, e.g. by hunting or creating shelters.

Humans lived as hunters and gatherers.

Around 60,000–10,000 years ago human hunters were able to drive particular large species (such as large predators) to extinction.

Around 10,000–2000 years ago humans developed agriculture ("neolithic revolution"), however at different times in different areas of the globe. Agriculture led to increasing, though still localised clearing of vegetation and an overall increase of human populations.

People nurtured some species (e.g. cultivating them) and fought others (as weeds or nuisances). Populations of some species as well as their habitats collapsed regionally as a consequence of human cultivation or exploitation of nature (e.g. forests in Greece).

Together with agriculture, humans developed complex societies and cultures, with increasing hierarchies and social division, e.g. as a matter of division of labour and power.

With a steady increase in population and a gradual increase of knowledge and technology (with drawbacks now and then, e.g. after the collapse of the Roman Empire), humans became less and less dependent on nature while having increasing impact on the land.

With the beginning of the "great geographical discoveries" in the 15th century, European practices and influences were brought as part of colonial expansion to most parts of the earth.

Around 1800 technology advances ("industrial revolution") led to strong population growth, accumulation of wealth, but also increasing inequalities in society, deteriorating health of city people, as well as increasing and largely unrestricted use of non-human nature, the latter perceived mainly as resource.

Stage 5: The 19th century, especially the end of the century, saw a crisis both in societal conditions as well as in the relations between humans and nature, perceived as matter of loss; the perception was fostered by an increasing knowledge and awareness of nature. Different groups of people focused on different phenomena as an expression of this crisis (e.g. cruelty against animals, species going extinct, traditional landscapes lost). Organised conservation efforts started.

Stage 3 and 4 repeated again: After WWII the industrial, economically and technologically driven influences of humans on nature strongly accelerated, and with it large-scale human impacts on non-human nature. This happened not the least due to the development of new persistent substances, such as plastic, and their release into the natural environment.

Stage 5: Between the 1960s and 1980s, awareness of a crisis of nature gained a new quality especially by more and more knowledge about ecology and about the consequences of human actions, consequences both for non-human nature as well as for most of humanity.

More and more species go extinct and habitats (e.g. rain forests) are being destroyed on a large scale.

Conservation efforts steadily increase.

Stage 6: Anticipating future events and providing "visions of nature". Alternative visions prevail: (a) humanity realises that strong efforts for safeguarding nature and stopping species extinctions are needed, especially through fundamental changes of human–nature relationships, in terms of a respectful coexistence not only with other humans but also with non-living nature *or* (b) continuing the current course of human interactions with nature leads to a collapse of both natural and social systems and an impoverishment of both human and non-human life.

Stage 7: Humans are able to come back to good relationships with nature by different means (a), *or* experience a collapse of civilisation as we know it (b).

What can be seen is that already this fabula is a selection of events. I left out, for instance, wars, or major catastrophes that happened (e.g. population crashes due to volcanic eruptions, cold periods such as the ice ages, or similar events). Of course, it is impossible to list everything that happened, no matter what scale (global, regional, local) is addressed. One *must* make a selection of things, in this case those that appeared to me the major events moulding human–nature relationships, even though other events – directly or indirectly – might also contribute. By making these choices, some kind of evaluation is done, namely of which known events are significant. Locally, there are major deviations, in the course of important events, when, e.g., people living in African rainforests, or Australian aborigines, did not develop agriculture in the sense of the neolithic revolution. The fabula described is also narrated largely from a Western perspective and would look different if it was assembled by, say, the descendants of the Maya, native people of North America, the Maasai in East Africa, or the aborigines in Australia.

From fabula to story

If it comes to filling in the details, to come closer to a real story, the same fabula is told with highly divergent selections and characterisations of actors and with different evaluations of events. In contrast to a fabula, the story is not bound to a strict chronological sequence. Actual stories often start with stage 5, the perceived crisis.

BOX 3.4 THE ANTHROPOCENE CONCEPT

The "Anthropocene" is not inherently a conservation concept. More to the contrary, its use is often strongly detested by conservationists. The concept is, in any case, highly relevant for conservation and for ideas about human–nature relationships.

The Anthropocene concept in its current form was introduced and popularised by ecologist Eugene Stoermer and atmospheric chemist Paul Crutzen (Crutzen and Stoermer 2000, Crutzen 2002). Their argument was that, due to their enormous influence on all parts of the earth system, humans now had become a dominant force on the planet, both in ecological as well as in geological terms. Therefore, they claimed, the geological period of the Holocene was coming to an end, being followed by a new geological period which they termed "Anthropocene". This proposal aroused intense debates, and meanwhile there are ongoing efforts to get a formal recognition of the Anthropocene as a new geologic epoch from the International Commission on Stratigraphy and the International Union of Geological Sciences.

Necessary for such a formal recognition of an epoch is inter alia evidence of a clear signature in the geological record, traditionally, e.g., by specific pollen records, new groups of fossils or the disappearance of previous ones, or the composition of mineral deposits (Lewis and Maslin 2015). Several candidates for the potential start of a formal Anthropocene Epoch have been proposed. Thornton and Thornton (2015) extracted four different ideas from the literature (see Figure 3.3), starting from the human-caused extinction of megafauna via the beginning of agriculture in the neolithic, the industrial revolution of the 19th

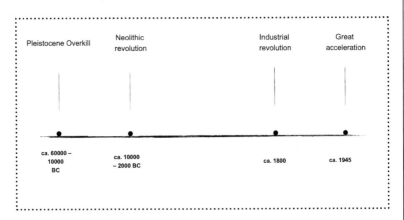

FIGURE 3.3 Different potential starting points of the Anthropocene. Figure by Kurt Jax, on the basis of data from Thornton and Thornton (2015).

century, up to the "Great Acceleration" of human activities after WWII, the latter signified inter alia by the appearance of new substances such as plastic or new persistent organic substances; Lewis and Maslin (2015) even list nine possible starting dates. All of these presumed starting points of the Anthropocene have been considered crisis points by some conservationists, i.e. points where a crisis of nature – or better: human–nature relationships – occurred or began. All of the previous states of nature, and the human–nature relationships associated to them, have been declared as desirable reference states for conservation by one or the other author.

Even more interesting for conservation than discussions about its precise starting point are the different interpretations of what the Anthropocene means for humans and their relationships with nature, expressed by different narratives (Bonneuil 2015, Thornton and Thornton 2015, Görg 2016). Some of these narratives see humans now in the "driver's seat" and postulate that we have to use our technical abilities to fix the damage done to nature (if necessary at all). Others perceive of it as a warning, lament the condition brought about, and demand a restriction of human activities to avoid apocalyptic outcomes of the human impact on nature. Yet others see the Anthropocene as a chance for humans to perceive themselves as a part of nature and to develop new relationships with nature. Common to most narratives appears to be the idea of an increased responsibility of humans with respect to nature, coming with the awareness of humans now being a dominating force on the planet.

In connection with this, some scholars ask – and doubt – if the "Anthropocene" is really something brought about *by humans collectively* (as a species, as the *anthropos*), and as to whether it affects all humans equally, bringing in matters of justice and the accountability of specific groups of humans (e.g. Chakrabarty 2009, Malm and Hornborg 2014; see also Section 5.2). In this context, the term "Anthropocene" has been criticised as a misnomer and several alternative terms have been suggested (e.g. Brondizio et al. 2016, Büscher and Flechter 2020). One suggestion has been to designate the current epoch instead as the "Capitalocene", thereby emphasising that it was a specific type of economy-driven activity of Western people that is responsible for and dominating the current state of the world.

The further consequences of the Anthropocene concept for conservation – if the concept is not outright rejected – are still under discussion (e.g. Wuerthner et al. 2014, Corlett 2015, Lorimer 2015, chapter 3 in Van Dyke and Lamb 2020).

The timing of the crisis, at least the one that is seen as decisive for the problematic current human–nature relationships, differs strongly depending on the specific story told, and there can be differences even with respect to whether something is seen as a crisis or not. Thus, the transition of humans towards farming (neolithic revolution) is evaluated by most nature conservation stories as at first positively, as a step towards more secure food provision, as part of an emancipation from the daily need to hunt and gather with all its vagaries. On the other hand, it has been perceived as already being the beginning of the ecological "fall of (hu)mankind", when humans began altering their environment, e.g. by clearing forests and also altering fauna and flora for their needs. For some conservationists the major crisis did not occur when agriculture started (with humans still having a good relationship to nature then) but only when agriculture became "industrialised", aided by machines and becoming large scale in the 19th century. The temporally earlier stages, then, are used to explain both what led to the perceived crisis as well as to find reference states of good human–nature relationships and as a good state of nature. The common evolutionary origin of humans and non-human organisms is, e.g., turned into an argument of their strong relationship and kinship – with responsibility ensuing, as in human kinship relations. Different stories were also told in the beginning of conservation, differing, e.g., in what was seen as the major non-human actants to be considered (individual organisms, particular species groups, landscapes, etc.) and also those actants perceived as causing the crisis (e.g. rich people, hunters, tourists, industry, etc.). The explanation is often linked to a justification for the future outlook (stages 6 and 7). Thus the idea of going back to "wild" nature without humans is sometimes justified both by a particular characterisation/role of humans in the crisis ("a nuisance to nature") as well as characterisation of the other major actant (nature) as autonomous, self-sustaining, and "knowing best" (see Section 3.3). It should be noted that also conservationists and conservation biologists can be and often are actants in narratives on human–nature relationships.

Specific conservation narratives (and their components) can be of different origins: some are routed in deep convictions and values, sometimes are of an "archetypical" character (e.g. nature as benign and harmonious), some are primarily "constructed" to convince other people (e.g. the economic value of nature), even if those putting them forward may be motivated by other convictions. They can also be more general and global (such as those relating to the links between biodiversity and climate change) or quite place-specific, especially when relational values come into play.

In the following, I will describe more specific narratives, as related to conservation concepts. The type of narrative which I will elaborate below are still a kind of meta-narratives, and ideal types, where actants are mostly characterised as types or roles and not as individuals, as time- and place-specific entities. It is, however, this middle level that is of special interest for the purpose of this book. More specific stories with individual human characters or groups, specific non-human objects, either individual organisms or collective entities, exist, but they will be only referred to as examples.

3.3 Major narratives in conservation concepts

My aim was not to describe and analyse detailed single stories of human–nature relationships in conservation, but to extract a limited number of conservation narratives, covering the most important ideas as found in conservation concepts. For this purpose, I used a partly deductive, partly inductive approach. I started from my own knowledge and from scanning the literature of what I perceive as "major conservation concepts", i.e. those concepts which are not in the first place technical, but which stand for programmatic approaches to conservation, such as wilderness, ecosystem management, or biodiversity (see also Section 3.4). In spite of the internal diversity of implicit human–nature relationships in these major concepts, I tried to carve out a limited number of idealised major narratives. To do this, I used the elements of the scheme for narrative analysis described earlier. Also, I looked for narratives told in the conservation literature and those used to communicate conservation issues to policymakers or the general public. I then went back to the concepts themselves, to see if there might be some other concepts embracing conservation narratives that I may have missed before; if necessary, I included them.

Six and a half major narratives of human–nature relationships in conservation and conservation biology

I will distinguish six major narratives here, and in addition a seventh one, which usually builds on these others. These narratives are not meant to be exhaustive nor always mutually exclusive, but, to my knowledge, they cover the most influential currents *within conservation* and *conservation biology*. There are definitely many more narratives as embraced by the broader public (see, e.g. the "counter narratives" of Louder and Wyborn 2020).

As major narratives describing human–nature relationships in conservation, I suggest the following:

- Humans apart from nature: guarding unspoilt nature
- Caring for other living beings (and natural features) is part of a good human life
- Safeguarding necessary and useful nature
- Being at home in nature
- Steering nature in a responsible way
- Gardening nature in novel ways
- [Making nature whole again]

In principle, a huge number of more differentiated narratives and types of human–nature relationship could be created. But it would not help for structuring the field of conservation concepts to present a long list here. The task is to find a middle ground between too simple and too complex. I will, however, for some narratives, point at possible sub-narratives, or branchings, as I call them.

The different human–nature relationships are not mutually exclusive. In part, several of them will be embraced by the one and the same person – with or without tensions and internal contradictions; narratives are not statements to be judged by the criteria of classical logic. They also involve idealisations and thus not necessarily depend on scientific "facts" to be relevant, forceful, and accepted – even though in the end they must have some substantial relation to (at least a potential) "reality".

In the following, each of the human–nature relationships will be described at first by a short overarching narrative description. It will then be elaborated further, inter alia by referring to conservation concepts that are representative for the respective narrative. The descriptions may appear as very coarse depictions of the narratives; they may even appear as creating caricatures or strawmen, as narratives told as by those objecting the respective idea. But it is necessary to emphasise and even exaggerate some characteristics of the concepts to provide clearer contours of the respective human–nature relationships.[6]

The major narratives will be characterised by the following features taken from the masterfabula developed earlier:

- What are the major actants?
- What are the considered properties of nature as a whole, or of more specific non-humans actants?
- What is the perceived *current* role of humans with respect to non-human nature?
- When did a crisis occur?
- What is the *desired* role of humans with respect to non-human nature?

I will name one or two paradigmatic conservation concepts for each narrative. There is, however, not a clear 1:1 correspondence between the narrative and the paradigmatic conservation concepts named. As will be shown in more detail in Section 3.4, many conservation concepts are so broad that, due to different varieties or interpretations of the same, they refer to several of the narratives. Finally I have selected one short story (or part of it) from the conservation literature in which the respective narrative is embodied. A comparative overview of the characteristics of all major narratives is provided in Table 3.1.

Narrative: *Humans apart from nature: guarding unspoilt nature*

Nature developed independently from humans. When humans appeared on the stage and started to transform nature, this caused a crisis to a previously untouched, pristine nature. Activities of humans are detrimental to nature. Nature is autonomous. Left on its own, it is robust and self-sustaining ("nature knows best"), sometimes even assumed of being in or tending towards equilibrium. However, nature becomes increasingly fragile the more humans interfere. Following this, humans in general are detrimental to nature and should protect it by leaving it on its own ("let nature take its course"). Humans should be (external) guardians or at best guests and then tread softly.

Paradigmatic concept: wilderness

This narrative is quite an old one in Western conservation. As described in Section 2.2, it was already part of colonial conservation efforts, in which "true" nature was seen as nature without any human interference, even without interference of "primitive" indigenous people. Also in many national parks this tradition is quite strong, as seen by the eviction of people from them, indigenous people (Spence 1999) but also others. For instance, when Shenandoah National Park in Virginia was created in the 1930s, some 500 families of white settlers were displaced from the area.[7] Also, the US Wilderness Act from 1964, e.g., aims to protect areas "where the earth and its community of life are untrammeled by man, where man himself is a visitor who does not remain" (see Section 3.4).

Many, if not most, ideas of "pristine nature" pursue this strong separation between humans and nature. Even though it was frequently seen as an act of cultural, "civilised" behaviour (Gissibl et al. 2012) to set aside areas completely free from humans, it was clearly setting nature and people, nature and culture, apart as two distinctive spheres. This relationship was – and often still is – also part of scientific ideas about nature, meaning that only nature *without* humans is a proper subject of ecological research, or is at least the indispensable baseline for studying and teaching how nature (or ecosystems, in scientific parlance) works (Inkpen 2017).

Even though strictly separating humans and nature, this relation is not one of disrespect for nature and its parts. To the contrary, it may even be seen as moral duty of humans to not interfere with nature, nature being valued "for itself". The "autonomy" of nature, i.e. the property of nature as "free" and existing on its own, is valued. This idea is often connected to notion of self-regulation (but see Hettinger 2005).

Modifications or branchings of this human–nature relationship have considered "primitive" or "savage" humans, living as hunters as gatherers, as part of unspoilt nature, even in need of protection as part of the wilderness. Thus, early proponents of a wilderness idea (Spence 1999 mentions especially George Catlin) wrote about an "Indian wilderness" in parts of the 19th-century USA, and in the early 20th century some European colonial powers, e.g. the Netherlands and Belgium, were pondering the idea of including endangered "natural peoples" in their conservation programs (Sysling 2015). Even towards the end of the 20th century, in the discussion of the "intactness" of Yellowstone National Park (or its northern range, respectively), Charles Kay saw "native Americans" as original "predators" of the area, and even its "ultimate keystone species" (Kay 1995, p. 114), whose eviction compromised the naturalness of the park. It is obvious that – even though expressed in politically correct terms ("native Americans") – the "naturalisation" of some "primitive" people and their cultures is highly problematic. In any case, the actants here become diversified, distinguishing "traditional" humans from "modern" humans.

Some people regard this kind human–nature relationship as strongly misanthropic, even though most of its proponents reject this allegation (see discussion in Keeling 2013). As with most of the other major narratives described later, there is not one unique motivation and one type of value behind this narrative and the relationship it describes, even though the idea of protecting nature "for its own sake" may predominate. I will discuss this in more detail in Section 3.4 with the example of the wilderness concept.

A story: Henry David Thoreau on his experience climbing Mount Ktaadn, Maine, having lost his hiking company.

> *It was vast, Titanic, and such as man never inhabits. Some part of the beholder, even some vital part, seems to escape through the loose grating of his ribs as he ascends. He is more lone than you can imagine. There is less of substantial thought and fair understanding in him than in the plains where men inhabit. His reason is dispersed and shadowy, more thin and subtle, like the air. Vast, Titanic, inhuman Nature has got him at disadvantage, caught him alone, and pilfers him of some of his divine faculty. She does not smile on him as in the plains. She seems to say sternly, why came ye here before your time? This ground is not prepared for you. Is it not enough that I smile in the valleys? I have never made this soil for thy feet, this air for thy breathing, these rocks for thy neighbors. I cannot pity nor fondle thee here, but forever relentlessly drive thee hence to where I am kind. Why seek me where I have not called thee, and then complain because you find me but a stepmother? Shouldst thou freeze or starve, or shudder thy life away, here is no shrine, nor altar, nor any access to my ear.*
> (Thoreau 1983/1864, p. 64)

Comment: This quote certainly does not mirror Thoreau's full idea(s) of human–nature relationships, which is much more complex. But it shows that nature for him was not simply something romantic, harmonious, but often strongly apart from humans, to be respected on its own (see discussion in Hoag 1982).

Narrative: *Caring for other living beings (and natural features) is part of a good human life*

Humans have always used nature, but at some point they endangered it in terms of mistreating individuals or extinguishing species or other natural features. It is thus not nature as whole but specific features of the same, especially species, to which humans relate and which matter. The losses of these features had negative effects on humans as well, in terms of affecting their feeling of what it means to be really human. A crisis in this respect occurred during the 19th century, even though perception of losses of specific features and even singular species extinctions have already been occurred earlier. To be truly human (and civilised), humans have to care for other species and see them as co-inhabitants of the world. They also have to care for their natural heritage. Humans are here regarded as responsible for protecting those items, before they disappear as a result of unheeding use of non-human nature. Nature is considered not completely apart from humans, but also not on equal footing; humans have a special role due to morality and reason. Living a good and really human life requires an attitude of care towards nature. Nature as a whole is seen as rather robust, but the continuing pressure of humans has endangered its fabric more and more. Humans have a responsibility to save nature (also in their own interest), as guardians, humble co-inhabitants, or heirs (stewards).

Paradigmatic concepts: biodiversity, natural monuments

Much of traditional conservation is based on this narrative and the human–nature relationship it describes. Preserving species as well as particular (rare or spectacular) natural features goes also back to the beginnings of Western conservation (see Section 2.2). This narrative is a rather broad one with many branchings. The specific justifications and expressions what makes up a real human life in relation to nature are diverse, ranging from avoidance of cruelty (as a matter of civilised humans), towards felt respect towards all natural beings, the assumed intrinsic values of the same, eudaimonistic reasons for protecting species, valuing diversity through issues of natural and cultural heritage. Even though the justifications for protecting nature may vary according to the respective relation of humans to nature, the core stays the same. Likewise, the objects of conservation vary considerably, from individual specimens of animals or plants, through species and even larger "natural monuments". The common core is that the relations are not based on functional considerations or on the value of the whole (eco)system. In contrast to such a more functional view of nature (e.g. narrative "Safeguarding necessary and useful nature", below), this narrative comes close to what Jax and Heink (2015) described as "item protection". It is about very specific, often irreplaceable items, independently of the functional roles they may play within ecosystems, the human economy, or nature as a whole.

At least three subbranches may be discerned:

Branching

a Animal protection
b Species protection
c Natural and combined biological and cultural heritage

The branches, different directions of the narrative, differ with respect to the objects that should mainly be preserved. The wish to preserve specific parts of nature is one connected to an attitude of *care* and *responsibility*, where the responsibility may be towards the specific objects themselves (intrinsic value), towards God, or towards society and humanity (including future generations).[8] In relation to Aldo Leopold, for instance, Coates (2015, p. 287) states:

> For Leopold, the preservation of endangered species was a marker of the civilized state of a national community and a measure of the moral progress of the human species as a whole.

The different branches of this narrative may nevertheless be in some tension, as manifested in the discourse of protecting species vs. protecting individuals – unless one sees issues of animal welfare as being completely outside of the realm of nature conservation (see the case study on the Oostvaardersplassen in Section 3.1).

BOX 3.5 ON THE DIFFICULTY TO JUSTIFY THE PROTECTION OF SPECIES AS SUCH

The extinction of species is of major concern for conservationists, something to be almost absolutely avoided and even seen as a moral wrong (Cafaro and Primack 2014). However, what does conservation really protect, when "species" are protected, and why? In practical terms, as a pragmatic object, it is not species as an abstract category that are protected but instead populations of particular species – by allowing their individuals to thrive and reproduce. The aim is to avoid the extinction of species as historical lineages which, once gone, are lost forever. While there can be a number of ethical arguments for protecting individuals (e.g. in terms of avoiding suffering or harm to individuals), it is rather difficult to argue for the value of *species* as objects for moral consideration. One may protect species for their utilitarian values, the functional roles their members play for humans. But that excludes all those species for which no functional role, no practical use to humans, is apparent; also the functional roles of many species may be filled by other species. The idea that an abstract thing such as species has intrinsic value is likewise contested among philosophers. Species cannot suffer and have no interests, only individuals can. The most convincing and soundest arguments for protecting really *species* therefore come from a eudaemonic perspective (see Section 3.1) and the question what it means to be human. McCord (2012) puts forward our essential ability of intellectual curiosity and wonder as a main argument, especially as related to the complexity, history, and evolutionary uniqueness of each species, beyond aesthetics and practical needs. Contributing to or tolerating the human-driven extinction of species deprives us of the future ability to experience miracle, wonder, and curiosity, which each species bestows on us, a thing that is not only of individual but of societal interest. Another, similar argument in this direction is the issue of personal identity related to parts of nature, as discussed earlier. This line of argument should be kept in mind when it comes to justify the protection of species as such.

A story: Edward McCord on discovering natural wonders.

> *Did you notice when the natural world first grabbed your attention? When you realized with all your sensibility that we are not alone – that our world teems with*

birds, bugs, reptiles, and other creatures that are certainly not human but are certainly also not unrelated to us? It is a stunning revelation and a memorable moment in a human life. It begins a journey that leads to frontiers of self-discovery and endless interplay between the life within us and the living world without.

We come from diverse homelands and childhoods that afford us varying engagement with the natural world. For my part, I grew up in North Florida, and I lost myself with friends in outdoor adventures as a boy. Ours was a land of salt marshes and fiddler crabs, of black lagoons and dark ravines, with watery sinkholes with walls of limestone and ferns, of crystalline rivers winding through glades of cypress and Spanish moss […].

Every child at some time surrenders, even if fleetingly, to the fascination of a living thing – a sparrow, a tadpole, a porcupine, a mushroom. Humans surely have roused the wonders of existence since time immemorial. But there is a new wind blowing among youth today. They are attracted in great numbers to courses in "environmental studies" that were completely unknown a few decades ago.[…]

These studies are surely ultimately and most relevantly about defining our human identity on earth.

(McCord 2012, p. ix ff)

Narrative: *Safeguarding necessary and useful nature*

Humans are dependent on nature or at least on selected objects and processes thereof. Humans are undervaluing nature and overusing it. Nature is considered to be rather robust when managed correctly and wisely. Humans are rather apart from nature; their role is the user of nature's goods and services. At first unproblematic and the basis of human life, crises occurred when human population strongly increased and at the same time humans massively extended their ability to extract resources from nature, up to the point where they endanger their own survival or at least the sustained flow of goods and services. In most cases the start of the crisis is seen as occurring with industrialisation (not the least in agriculture) by the middle of the 19th century and even more after the "Great Acceleration" after WWII. So humans should move from reckless users to using nature prudently and in a sustainable way and protect those parts of nature on which their survival and well-being depends.

Paradigmatic concept: ecosystem services

Functionality, with respect to human survival and well-being, is paramount for the question of what in nature is worthy to be saved and managed. This narrative and human–nature relationship is at the core of resource management, manifested, e.g., from early forest management to today's idea about optimising the use of ecosystem services and landscape functions in continental Europe. It has to be said that this narrative is often one used by conservationists in a highly strategic manner, aiming in fact at much more than mere resource protection, as, e.g., debates about relations between ecosystem services and biodiversity show (e.g. Mace et al. 2012, Deliège and Neuteleers 2015, Jax and Heink 2015).

Branching

This narrative has various varieties and related emphases. These can refer to the following:

a (Mere) survival
b Purely economic usefulness
c Human well-being in a broader sense (including cultural benefits of nature)

For more on this narrative, and in particular the dynamics and complexity of the ecosystem services concept, see Section 3.4.

A story: Gretchen Daily's story about the requirements to live on the moon

> *One way to appreciate the nature and value of ecosystem services […] is to imagine trying to set up a happy, day-to-day life on the moon. Assume for the sake of argument that the moon miraculously already had some of the basic conditions for supporting human life, such as an atmosphere and climate similar to those on earth. After inviting your best friends and packing your prized possessions, a BBQ grill, and some do-it-yourself books, the big question would be, Which of earth's millions of species do you need to take with you?*
>
> *Tackling the problem systematically, you could first choose from among all the species exploited directly for food, drink, spice, fiber and timber, pharmaceuticals, industrial products (such as waxes, lac, rubber, and oils), and so on. Even being selective, this list could amount to hundreds or even several thousand species. The space ship would be filling up before you'd even begun adding the species crucial to supporting those at the top of your list. Which are these unsung heroes? No one knows which – nor even approximately how many – species are required to sustain human life. This means that rather than listing species directly, you would have to list instead the life support functions required by your lunar colony; then you could guess at the types and numbers of species required to perform each. At a bare minimum, the spaceship would have to carry species capable of supplying a whole suite of ecosystem services that earthlings take for granted.*
>
> (Daily 1997, p. 3)

Narrative: *Being at home in nature*

Humans are part of nature and jointly have created stable landscapes in which they interact in a harmonious way. Nature is home both to humans as well as to other species. Humans and other species, as well as the landscape setting that allows its thriving, are in close interaction and created harmonious landscapes patterns, benefitting both humans and non-human nature. Non-human nature is used but also cared for. Nature is shaped and perceived as malleable, but only within certain limits. Sometimes nature is even seen as being improved by human actions, e.g. in terms of higher diversity of habitats and species. In some cases, species (e.g. large predators, diseases, or late successional species) may also be contained or locally eradicated in the favour of being able to maintain the overall landscapes and its working. Culture and nature are seen as going hand in hand. Symbolic issues, such as the feeling of home or belonging, play a crucial role. This narrative has a place-specific emphasis, much less than a global one. Nature is seen largely as benevolent, partly as results of humans removing the most dangerous species. A crisis of this kind of human–nature relationship occurred mainly with the advent of industrial agriculture, large-scale resource exploitation, and, to a lesser degree, tourism, all of it considered to be starting in the 19th century. This crisis diverted humans from their roles of caretaker, partner, or cultivator to one of mere (unwise) resource user or even destroyer of established landscapes. The old roles should be re-established.

Paradigmatic concept: **cultural landscapes**

Mostly, such relations have developed for an extended historical time, on almost all continents. Relations are not only material but to a high degree also symbolical, e.g. relate to people's personal identity. However, expressions of this relationship differ inter alia with respect to how much a particular – mostly historical – state and/or type of interaction should be conserved (even restored) or as to whether the interactions between humans and nature should further develop within these landscapes, probably also changing their appearance. Should, for instance, new (exotic) species be part of them (as they have to some degree been in the past)? What counts as the sense of place by which people define their identity in relation and physical interaction with non-human nature? As a conservation narrative, these ideas have been present from the beginnings of conservation, but initially restricted to more densely settled landscapes in Europe.

A story: Rachel Carson's "A fable for tomorrow".

There was once a town in the heart of America where all life seemed to live in harmony with its surroundings. The town lay in the midst of a checkerboard of prosperous farms, with fields of grain and hillsides of orchards, where, in spring, white clouds of bloom drifted above the green fields. In autumn, oak and maple and birch set up a blaze of color that flamed and flickered across a backdrop of pines. The foxes barked in the hills and deer silently crossed the fields, half hidden in the mists of the fall mornings.

Along the roads, laurel, viburnum and alder, great ferns and wildflowers delighted the traveller's eye through much of the year. Even in winter the roadsides were places of beauty, where countless birds came to feed on the berries and the seed heads of the dried weeds rising above the snow. […] So

94 Analysing conservation concepts

it had been from the days many years ago when the first settlers raised their houses, sank their wells, and built their barns.

(Carson 1962, p. 13)

Comment: this famous and highly poetic quote is the opening passage of Rachel Carson's book *Silent Spring*, on the consequences of extensive pesticide use. It does, still quite unusual for American conservation-related texts, describe a *cultural landscape* as an ideal, not a tract of wilderness.

Narrative: *Steering nature in a responsible way*

Humans are members of a larger biotic community and of social-ecological systems and have to steer these in a responsible way. Humans ran into the ecological crisis, inter alia because they focused too much on the use and management of isolated resources (parts of nature). Nature in principle is robust but a highly complex system, linked to human (social) systems. Given this high complexity of nature and the knowledge gained by ecology and other sciences, humans have to manage not just isolated features of nature (populations, species, resources), but whole chunks of nature (ecosystems; "the land"), to maintain them healthy and functioning. We, as humans, now should steer these systems prudently, for our own sake and for that of nature. Humans are both members of a larger ecological community as well as stewards of nature.

Paradigmatic concepts: ecosystem management, social-ecological systems

This narrative is a rather recent one, in large part inspired and guided by ecological research. It takes a strong systems perspective, often with humans and their societies as part of larger social-ecological systems. Stewardship can be understood alternatively as a means to most efficiently manage nature or as an attitude of care towards the whole of nature (see Box 3.6). In contrast to "Safeguarding necessary and useful nature" or "Caring for other living beings (and natural features) …", it focuses on whole ecosystems, be they "natural" (e.g. wilderness areas) or ecosystems where humans and nature strongly interact. In contrast to "Being at home in nature", it contains fewer symbolic and emotional dimensions. The narrative of "Steering nature in a responsible way" and even the idea of seeing nature as being composed of social-ecological systems has been expressed, e.g. by Aldo Leopold (Meine 2020), where in fact it may overlap with the narrative of "Being at home in nature" (Callicott 2000a) and is often related to the concepts of "land health" or "ecosystem health" and also extends into areas without humans. In contrast to "Being at home in nature", the narrative has also a strong relation to a scientific and systems view. This may appear as a marginal difference, but I think we should heed it.

A story: Management of the St. Clair River by the Walpole Island First Nation

The St. Clair River is a 65-km-long river that flows south, draining Lake Huron into Lake St. Clair and forming part of the international boundary between

Ontario and Michigan. The river flows through a heavily urbanized and industrialized landscape. [...] The river branches into several channels near its mouth at Lake St. Clair, creating the Walpole Island Delta, which is home to Walpole Island First Nation [WIFN]. In the Anishnaabe language, the area is called Bkejwanong (where the waters divide). [...]

The people of Bkejwanong actively care for their waters in a number of ways, including environmental monitoring and research, government-to-government political activities at local, provincial and national levels, and caring for waters spiritually. This spiritual path sits at the center of all the community's water-related work and is led by a group of women of Bkejwanong known as Akii Kwe (Earth women). According to Anishnaabe traditions, women have the responsibility to care for water spiritually and to speak on behalf of water. [...]. The community recognizes that it is on a path of healing human community members following the traumatic impacts of colonialism, as well as healing the water, which is also viewed as a living member of the community or extended family.

While the tribe has been sustainably managing this ecosystem for thousands of years, land use change on Walpole Island and upstream pollutants are currently creating challenges to sustainable management. WIFN has worked to protect the river, which, as Anishnaabe people, they believe is an obligation. They are active in remediation and restoration of the St. Clair River in their territory, sitting on bi-national advisory committees and partnering with environmental groups and other communities.

(Fox et al. 2017, p. 526ff)

Comment: This is in fact a story from a recent conservation study, describing a project and existing practices related to the management of an area, specifically a river, both perceived as an ecosystem (in research) as well as a member of a community of humans and non-humans (spiritually). Scientific insights are linked with traditional knowledge and practices. This story also links to the narrative that I have labelled "Making nature whole again" (see later), with its references to healing nature and humans.

Narrative: *Gardening nature in novel ways*

There is almost no place on earth which is completely "natural" any more, in the sense of not impacted in some way by humans. Humans exert influence on all parts of nature, i.e. the non-human actants can be all individual parts of nature and the complexes they form (in scientific parlance: communities, ecosystems). Nature is perceived as robust and malleable, even though not without limits, and humans are the dominating force on earth, having created completely new ecological assemblages, with species and ecological conditions in new places and with new interactions. A crisis, if it is perceived at all, started when humans entered the Anthropocene (here industrialisation, mid-19th century)

> **BOX 3.6 STEWARDSHIP CONCEPTS**
>
> One desired role of humans in nature is often described as that of a steward, with stewardship denoting the attitude and practice connected to it. According to the Merriam-Webster Dictionary, a "steward" is "one employed in a large household or estate to manage domestic concerns (such as the supervision of servants, collection of rents, and keeping of accounts)". This is a meaning that goes back to the 12th century.[9] In this sense, a steward belongs to what he or she takes care of (the "household") but is not its owner. In a conservation context, stewardship thus means taking care of and managing nature in a responsible way. In the environmental and conservation realm, many qualifiers of stewardship can be found, such as animal stewardship, environmental stewardship, ecosystem stewardship, landscape stewardship, wilderness stewardship, and even earth and planetary stewardship (e.g. Norton 2002, Chapin et al. 2009, Hobbs et al. 2010, Rozzi et al. 2015, Bennett et. al. 2018).
>
> Some scholars interpret stewardship in terms of a prudent (resource) management. Chapin et al. (2009), e.g., state that "Ecosystem stewardship is an action-oriented framework intended to foster the social–ecological sustainability of a rapidly changing planet." (p. 241) and "its central goal is to sustain the capacity to provide ecosystem services that support human well-being under conditions of uncertainty and change" (ibid, p. 242). Others conceive of stewardship as the broader set of human relationships with nature and as a morally grounded caring attitude and practice (Rozzi 2015a, West et al. 2018). There are different understandings of how much humans must and should intervene in nature, and how much such stewardship aims at the well-being of humans or also, even mainly, at that of non-human beings. Some critiques see the idea of stewardship almost as hubris – as based on the assumption that humans know best what is good for nature. Others perceive it as a humble attitude towards nature, with an obligation of caring (Bakken 2009). The concept of stewardship has also been used in (and partly is rooted in) religious and spiritual traditions. Here, the responsibility for taking care of nature is neither to humans nor to nature itself but towards God or spirits.

and caused partly irreversible changes of previous ecological conditions and mixes of species (exotic species). It might be argued that this point was reached even earlier, when species were distributed by humans over larger distances, out of their native areas (i.e. in the neolithic). As humans cannot come back to a "pristine" nature, their role is that of a prudent tinkerer and should be that of a responsible, humble gardener, not of a designed garden, but of a "wild" garden.

Paradigmatic concept: novel ecosystems

This is perhaps the most recent narrative, guided by the perception of Earth being completely dominated by humans. Dominated does, however, not imply "controlled" in the sense of a "conscious or intelligent control" (Marris 2015, p. 44). It describes that nature is valued *also* in appearances which are not the "traditional" ones. The metaphor of the *wild* garden is used here to express that this narrative is different from that of cultural landscapes ("Being at home in nature"). It emphasises the middle ground between controlling nature (e.g. for providing ecosystem services, but also by cleansing a place from all exotic species) and a complete absence of any guidance. Thus, also uncommon, new types of ecosystems and communities are left "wild", with a mild and humble guidance, however, by which conservation values are advanced. The human gardener promotes the flourishing of non-human organisms, is open to surprises, unexpected developments, tinkers with them (a difference to the "Steering nature in a responsible way" narrative). The narrative thus moves beyond classical dichotomies of nature and culture. The garden metaphor and the narrative described here are highly contested among conservationists (e.g. Soulé 2013, articles in Wuerthner et al. 2014). For a thorough review of the use of the garden metaphor in conservation and restoration, see McEuen and Styles (2019).

A story: Emma Marris recounting the rise of a new ecosystem on an Atlantic island.

> *Ascension Island in the South Atlantic is famous among ecologists. Usually they tell tales about aggressive invader species bulldozing complex ecosystems. Here it was the other way around. Ascension started out as a monotonous plain of ferns; a visiting Darwin dismissed it in July 1836 as "very far inferior" to nearby St. Helena. Seven years later botanist Joseph Hooker recommended the island be improved by the importation of many new kinds of plants – what modern ecologists might see as a massive invasion. But this invasion resulted not in ecological meltdown but in the creation of a cloud forest composed of species from here and there, trapping mists, cycling nutrients, and surviving, generation after generation, all without having evolved together.*
>
> *According to ecology textbooks, millennia of coevolution are required to set up complex interactions between plants, animals, microorganisms, nutrients, water, and other components of ecosystems. The idea that a bunch of random plants from hither and yon could convincingly impersonate a real ecosystem rather than collapse into species-poor and poorly functioning wastelands has been markedly hard to swallow.*
>
> (Marris 2011, p. 111)

Narrative: *Making nature whole again*

Humans – in one or the other way – have failed on nature. Nature is thus seen as a whole whose health or integrity has been compromised by human activities. As *a forward-looking message the wounds inflicted on nature should be healed and the health of nature should be restored, as well as the relationship between humans and nature: sometimes, even the relations between humans, within their societies, should be healed by the process of restoration.*

Paradigmatic concept: restoration

This is a special type of narrative as it mostly builds on one of the other narratives described earlier and thus may be seen as a part of these, but, if used so, an important part. That means, the first steps of the complete narrative can come from different other narratives, different ideas about what part of nature or which human–nature relationship are in crisis.

Aldo Leopold (1993, p. 165), in a famous expression, spoke about nature in the 20th century as a "world of wounds", a condition which is recognised by ecologists and is to be healed. The notions of "repairing" or "healing" are explicit in the restoration literature, even in the titles of books and papers (Hall 2005, France 2008, Blignauth and Aronson 2020). They sometimes, at least implicitly, build on a feeling of shame or guilt for what humans did to nature, and a sense of responsibility to redress these impacts (e.g. Hertog and Turnhout 2018). The ideas of what to be made whole again, repaired or restored, however, vary a lot, as do the baselines (or reference conditions) to which nature should be brought back. Are ecosystems be repaired for means of better provision of ecosystem services, healed for the sake of species survival, restored to provide an idea of how nature was before Europeans arrived ("a vignette of primitive America", as postulated in 1963 for American national parks; Leopold 1963)? Healing and repairing is not limited to repairing nature but also can extend to the relationships between humans and nature themselves, and by doing this contribute indirectly to healing intra-human relationships (Fox et al. 2017). This is accompanied by a variety of specific terms and concepts, expressing these different goals, such as restoration, rehabilitation, or remediation (see Section 3.4).

A story: Robert France on restoring the Great Swamp in Cambridge/ Massachusetts

Gazing out the window at what was once the ecologically vibrant Great Swamp, I watch the runoff coursing off the impervious surfaces from the storm of a few hours before, and smell the toilet waste discharged through the antiquated combined sewer overflow system that characterizes many old cities such as Cambridge [Massachusetts]. And I think about how the nature of the all-but-dead wetland continues to force us to reflect upon its former namesake greatness. It was only a few years ago, for example, that the "storm of the century" flooded hundreds of homes when the wetland suddenly remembered its old boundary and the river its old course. Fish, some natives but unfortunately mostly exotics, were seen swimming on golf course fairways, birds paddled around immersed automobiles left stranded in parking lots, and plants continued to sprout from amongst the broken pavement in abandoned lots for months afterward. Nature obviously wanted to come back; but if its return was to be more than ephemeral, it needed guiding and stewardship by concerned humans.

Just as the Great Swamp is a striking lesson in ungratefulness, remorse and shame about how we have treated the wetlands that helped shape this country, the area is now on the cusp of becoming a signature example of dedication, compensation and hope in how to "restore" natural structures and functions in dense urban cores. "Restore" here is in quotation marks because any ameliorative and reparative actions taken towards returning a sense of naturalness to the area and a sense of wildness to its human inhabitants will never be an exact replica of both the ecological and ecocultural relationships that once existed in the area in pre-(European)contact times.

(France 2008, p. 3)

Comparing the major narratives

Table 3.1 gives a short overview of the six (seven) narratives along the features by which they were described. As said before, these narratives are neither exhaustive nor all mutually exclusive. Most individuals will not subscribe to only one of the narratives when it comes to characterising one's own relationship to nature. It is obvious that we all need non-human nature to survive and also for our well-being ("Safeguarding necessary and useful nature"). But that utilitarian view on nature does not preclude others, at least not in general. If we speak about specific areas or places, however, or about specific conservation measures, there can be mutually excluding narratives and human–nature relationships. Here, e.g., a view of "Nature apart from humans" will run into conflict with one of "Being at home in nature" – such as when it comes to protect particular management-dependent (semi-natural) areas in Europe or even Australia. Australian aboriginal people have managed the land, e.g. through the intentional use of fire, long before Europeans arrived on the continent (Kohen 2003; see also Section 4.3).

The narratives vary inter alia in the degree to which much humans (and specifically which groups of humans) are seen as a part of nature or not. Also, they differ in the importance that scientific thinking has in shaping them. This, e.g., makes a decisive difference between "Steering nature in a responsible way" (which is built heavily on ecological and systems thinking) and "Being at home in nature", which is not.

The different narratives do not map easily unto unique values dimensions. From the very beginning, there have, e.g., been multiple ideas why to protect particular species as part of a good life – based on a range of values: relational values, intrinsic values, and even to some degree utilitarian values. Even for a concept like "ecosystem services", where the narrative ("Safeguarding necessary and useful nature") seems to be clear and its associated values as well (instrumental ones), other values are often embraced. I will come back to this in Section 3.4.

TABLE 3.1 Major conservation narratives and their properties (see text)

	Humans apart from nature	Caring for other living beings and...	Safeguarding necessary and useful nature	Being at home in nature	Steering nature in a responsible way	Gardening nature in novel ways	Making nature whole again
Major actants	Humans, non-human nature as a whole	Humans, specific items, especially species, also individual specimens and remarkable natural features	Humans, selected components, and processes of ecosystems	Humans, non-human species, and abiotic boundary conditions, in interaction	Humans, ecological and social communities, entire ecosystems	Humans, all individual parts of nature	Humans, ecosystems, landscapes
Properties of nature	Autonomous. Robust, when left on its own, fragile when humans interfere	Autonomous, but rather passive; robust but under increasing pressure	Robust and rather resilient	Malleable but sometimes fragile	Robust, complex, linked to societal systems	Malleable, robust	Depends on previous human impact: resilient/robust or fragile
Role of humans (hitherto)	Humans are detrimental to nature	Destroyer of nature, careless user	(Sometimes unwise) User of nature's goods and services	Partly caretaker, partner, cultivator, partly destroyer of nature, unwise user	Unwise user	Dominating force on Earth	Unwise and unsustainable user, destroyer
Crisis	When humans started to transform nature	Not clear: Pleistocene or 18th/19th century	Industrialisation of life (agriculture/urbanisation) in 18th/19th century	Industrialisation of life (agriculture/urbanisation) in 18th/19th century	Industrialisation of life (agriculture/urbanisation) in 18th/19th century	If at all, when entering the Anthropocene	At various times, depending on specific case

Analysing conservation concepts 101

Desired role of humans	Guardian and occasional guest	Guardian, humble co-inhabitant, steward	Prudent user (sustainable)	Caretaker, partner, cultivator	Steward, member of a larger biotic community	Responsible gardener and tinkerer	Healer, restorer
Paradigmatic concept(s)	Wilderness	Biodiversity, natural monuments	Ecosystem services	Cultural landscapes	Ecosystem management	Novel ecosystems	Restoration

Overall, there is a variety of ideas about existing or desirable human–nature relationships in conservation concepts (see Flint et. al 2013, Muradian and Pascual 2018, Louder and Wyborn 2020), often implicitly. What is still reflected too rarely in conservation concepts, however, is the relation *between humans* (or different groups of humans) in the context of human–nature relationships, even though the issue plays an important role, e.g., in the debate about the "new conservation" (Section 2.8). As will be seen in Chapters 4 and 5, this is something to which more attention must be devoted, because our relations to other humans cannot (or at least *should* not!) be dealt with in isolation from our relations with nature (and vice versa). Narratives of human–nature relationships are helpful for comparing Western conservation concepts with ideas of human–nature relationships from non-Western societies (Chapter 4). More than that, they are a useful starting point for searching appropriate approaches towards conservation in specific settings (Chapter 5).

3.4 Conservation concepts: conceptual clusters and a matrix

This section will broaden the approach to conservation concepts, both by including a larger set of concepts as well as by bringing together the different analytical perspectives which were introduced in Section 3.1, by looking closer also on the objects and values embraced by the concepts and the strategical dimensions included in them. With respect to the major narratives developed in Section 3.3, it will emerge that the relation between narratives of human–nature relationships and specific conservation concepts, even for some of those named as paradigmatic concepts, are not as clear-cut as it may appear at first sight. That happens because conservation concepts are a kind of container: they are not static but change, develop, and diversify through time. Their continuous development shows that the narratives originally connected to them are contested. The narratives also change, and the resulting modified narratives may even contradict each other. The term and a certain core remain the same, but "under the hood", there are efforts to, e.g., account for criticism. This will be demonstrated in some more detail for the ecosystem services concept and related concepts.

I have structured the larger set of conservation concepts by assembling them into "conceptual clusters" of related terms and concepts. The titles of these clusters correspond to those concepts that were named as paradigmatic for the major narratives of human–nature relationship elaborated in Section 3.3. This selection is based on the assumption that these paradigmatic concepts are at the same time also key words for programmatic approaches to conservation. In addition, I will discuss some other concepts important to conservation that I do not see as independent programmatic approaches but which are *supportive* to several of the major concepts. These are concepts that characterise desired properties of target objects, e.g. stability, functioning, resilience.

In the second part of this section, the analysis of conservation concepts will finally be summarised in the form of several comparative tables built on the different analytical perspectives, together forming a kind of matrix for analysing conservation concepts.

Conceptual clusters of conservation concepts

A conceptual cluster, as understood in this book, assembles concepts with common properties in terms of their epistemology, their meaning, and function in conservation, and the phenomena they describe. Such concepts may in some cases be synonyms (different terms for roughly the same concept) or content-wise related ones, such as ecosystem services, natural capital, and human well-being as part of what I describe as the "ecosystem services cluster". Astrid Schwarz and I used this approach to structure the multitude of ecological concepts, but it can also be applied usefully to the context of conservation biology (see Jax and Schwarz 2011, p. 13ff, for details on the idea of conceptual clusters).

Please note that there may be many more terms for each cluster than listed here, not the least technical ones, relating to assessment and management tools. Also, in a few cases, the same term may be part of more than one cluster. Moreover, some terms, in particular, ecosystem services or biodiversity can both refer to a programmatic approach, but also to specific objects (e.g. quantified ecosystem services such as timber, or biodiversity understood as species numbers). The latter will become relevant in the tables at the end of this chapter.

Cluster "Ecosystem services"

Cluster: ecosystem services (including subcategories such as cultural services), natural capital, human well-being, nature's services, functions of nature, nature's contribution to people (NCPs); green infrastructure; nature-based solutions, payments for ecosystem services

This cluster will be elaborated in more detail than the others, to exemplify the dynamics and changes of conservation concepts through time. The whole cluster is structured around the idea of the benefits or contributions humans derive from nature and the strategic aim to make the values of nature to humans visible – or even part of decision and management processes. It assembles a rather large number of concepts (more may be added to the list above) which are similar, or which are related to this idea, some as simply supporting it or being methods to make use of it (e.g. payments for ecosystem services, green infrastructure).

The concept of ecosystem services was developed with a clear applied and then even *strategic* purpose, namely at first to demonstrate and then also quantify which value nature has to humans, also beyond the traditional market values of tradable natural resources like timber or grain. It was introduced originally mainly from a conservation perspective and an economic perspective.

In conservation, Ehrlich and Ehrlich (1981, p. 6) introduced the concept in a much narrower sense than used today. For them, the notion of ecosystem services, a new term which they introduced in their book, was one of the four major arguments for the conservation of species, the others being ethical (compassion), aesthetic, and economic. Ecosystem services were the "free services" provided by species in keeping the Earth's ecosystems running, and thus the human life support systems. The Ehrlichs' also connected this argument explicitly with their famous "rivet popper" metaphor (see Section 3.2) with which their book starts. The idea of ecosystem services as described here was in no way an economic one. Such an economic perspective on the services of nature to humans, however, was already put forward by Westman (1977), who spoke of "nature's services", and most prominently, 20 years later by the seminal paper of Robert Costanza and colleagues (1997).

With the Millennium Ecosystem Assessment (MA, 2005) the concept entered the political arena as a tool to assess the state and trends of the Earth's ecosystems. It also intensified further research on ecosystem services. Since then the concept has been used in various scientific and political contexts, with purposes ranging from a merely "didactic" tool for demonstrating the benefits of nature, to application for assessment, planning, and for approaching environmental conflicts. The MA itself did not perform an economic valuation of ecosystem services but in its major graph (see Figure 3.4) referred to multiple benefits of biodiversity and ecosystems for

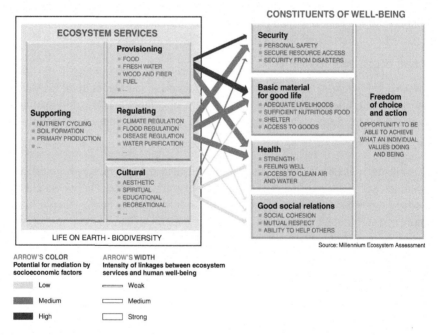

FIGURE 3.4 The MA scheme showing the relation of ecosystem services to human well-being. The MA classified ecosystem services into four types: supporting, provisioning, regulating, and cultural. Graphics: Millennium Ecosystem Assessment, 2005.

human well-being. The idea of human well-being, described there and visualised in the graph, reached much beyond any simple notions of well-being like GDP (economic well-being) but characterised well-being along several dimensions derived from an empirical study named "Voices of the poor" (Narayan et al. 2000). Beyond "basic material for good life" these were also security, health, good social relations, and freedom of choice and action. Well-being thus became a central element of the ecosystem services approach.

The perception of the ecosystem services concept as being focused (only) on economic values, however, was intensified by the use of additional concepts such as "natural capital" (Costanza and Daly 1992) or "total economic value". This language had its effect in the circles for which it was targeted: it is meanwhile used broadly in many administrations, is part of political environmental strategies, and is even included in the name of the "World Biodiversity Council" IPBES. However, many conservationists were highly appalled by this economic focus and felt that it was counterproductive to their conservation ideals (e.g. Redford and Adams 2009), that it might lead to a commodification and "selling out" of nature (Gómez-Baggethun and Ruiz-Pérez 2011). These concerns led to intense debates within the community of researchers and users committed to the ecosystem services concept and to partial refinements of the concept.

The *strategic use* of the ecosystem services for conservation led into a conundrum for many conservationists. The challenge is how to maintain the strategic powers of economic arguments while at the same time avoiding the fate of Goethe's sorcerer's apprentice, i.e. being unable to control the powers set free, in terms of letting a merely utilitarian perspective of nature taking over completely in the public perception of conservation. This means to object the view that conservation is *only* about economic reasoning or at least only about a human–nature relationship of "Safeguarding necessary and useful nature". In fact, most conservationists known to me that use an ecosystem services approach also use it – even first and foremost – with the intention to foster the protection of species and other facets of nature, beyond mere utilitarian thinking. These conservationists and conservation scientists at the same time also embrace other relations to nature (such as "Caring for other living beings…"), without denying that humans also need nature for survival and well-being.

Several remedies have been discussed and developed within the ecosystem services community for countering extreme utilitarian connotations of the concept, partly reshaping the concept. On the one hand, the links, even synergies between biodiversity and ecosystem services were increasingly emphasised. This was not the least seen as a means to avoid the neglect of biodiversity in the conservation debates and to emphasise other human–nature relationships that were seen as connected to it, e.g. the perceived intrinsic value of species, or at least the relational values of biodiversity (see Mace et al. 2012, Jax and Heink 2015 for this debate). Second, valuation schemes were strongly refined, to broaden the conceptual scope for valuing nature, to which the ecosystem services approach is closely linked.

This involves moving away from a valuation of nature that tries to make all values comparable by expressing them in a single "currency" (usually money), towards a more pluralistic view of value that incorporates intangible, cultural, and relational values (Kenter et al. 2015, Arias-Arévalo et al. 2018, Himes and Muraca 2018). A third approach reacting to critique and towards "cleansing" the ecosystem services concept from its economic connotations was the more recent effort to complement or substitute the concept by a similar, albeit broader one, namely that of "nature's contribution to people" (NCPs) (Pascual et al. 2017, Diaz et al. 2018). This happened in the context of debates in the IPBES (see also case study in Section 4.4). "Nature's contribution to people" were defined as:

> [a]ll the positive contributions or benefits, and occasionally negative contributions, losses or detriments, that people obtain from nature. It resonates with the use of the term ecosystem services, and goes further by explicitly embracing concepts associated with other worldviews on human–nature relations and knowledge systems.
> *(Pascual et al. 2017, p. 15)*

Especially the relational value concept was highlighted explicitly here, added to merely instrumental values.

There is, however, considerable disagreement over the extent to which NCPs help to address the conservationists' critiques. It was argued inter alia that, on the semantic level, NCP still depicted "the relation between nature and people as one-way and the value of nature as instrumental (as a provider of benefits), masking human agency and broader values" (Kenter 2018, p. 40). Muradian and Gómez-Baggethun (2021, p. 1) state that both concepts embrace "dualistic, anthropocentric and utilitarian representation of human-nature relationships", which they see as a major reason for the current environmental crisis; in consequence, they postulate an alternative human–nature relationship, with a morality of care instead of utility.

What I wanted to demonstrate with the elaboration of the ecosystem services concept and its history are several points. First of all, it demonstrates how dynamic conservation concepts are. There is not one "correct" understanding of a concept nor do the meanings of the terms remain the same. The research communities in a particular field (here ecosystem services) continuously develop concepts further, for the purpose of better applying them, but also in response to external as well internal critique.

Second, the values as well as of the ideas of human–nature relationships underlying specific concepts are not as clear-cut as it appears at first sight and as opponents of a concept often allege. However, in terms of the concept of ecosystem services specifically, my impression is that while the ecosystem services concept is in fact much broader than imposing an economic rationale on humans' relationship with nature, it will never get rid of the economic connotations related to term "services" (let alone natural *capital*), no matter how much explanations will be given.

Concepts of the ecosystem services cluster have had enormous influence on national and international policy arenas and were partly even developed for that purpose (e.g. the term "nature-based solutions"; see Nesshöver et al. 2017). The question of how conducive this influence has been for conservation is highly contested; descriptions of ecosystem services and related approaches have, however, sometimes been oversimplified and distorted, neglecting the complex debates surrounding them (see also case study in Section 2.8).

In principle, the ecosystem services concept has been a modern way to link humans and nature instead of setting them strictly apart, as it emphasises human dependency on nature and the various ways in which it contributes to human well-being. It thus can be seen as a progress for conservation. As stated earlier, there are, however, also other points of view, namely that ecosystem services are instead favouring a human–nature dualism by largely providing an unidirectorial and utilitarian perspective on human relations to nature, excluding relations of care and reciprocity (e.g. Muradian and Gómez-Baggethun 2021).

As described the spectrum of **values** of nature which proponents of ecosystem services and related concepts try to cover is broader than just instrumental values, even though there is debate what these concepts can and *should* cover, e.g. when it comes to conceptualise "cultural ecosystem services" (see. Kirchhoff 2019). This links also to the **objects** which are addressed by concepts from the cluster. The explicit ultimate object is certainly humans and their well-being. The pragmatic objects then are diverse, namely *selected* ecosystem processes and objects that are of use to humans. Organisms, however, are addressed less in their character as individuals or species but mostly in their functional roles, e.g. as promoting climate regulation or providing food (see Jax and Heink 2005). Understood as a **strategic** approach to promoting nature conservation (preserving ecosystem services and at the same time biodiversity, or even using the former as a means to save the latter), the ultimate objects may also be ecosystems or species.

While the "Safeguarding necessary and useful nature" is clearly the dominant **narrative** behind the concepts of this cluster, some of them, on behalf of the strategic use of the concepts, can also have important relations to other human–nature relationships, such as "Being at home in nature" for cultural ecosystem services.

Cluster "Biodiversity"

Cluster: biodiversity, species diversity (including varieties such as functional diversity and phylogenetic diversity), genetic diversity, ecosystem diversity, habitat diversity, biocultural diversity, biophilia, endangered species, extinction, natural heritage, natural monument, charismatic species

The *term* "biodiversity" is a neologism that was coined in 1986 (Takacs 1996; see also Section 5.1). From the onset it was not just a term to denote a scientific measurable object, but also a strategic term, to foster conservation. Even more, the word "biodiversity" designates a whole research and conservation programme, most clearly visible in the use of "biodiversity" in policy, e.g. the Convention on Biodiversity, the EU Biodiversity Strategy, or the IPBES. "Biodiversity conservation" often is used almost synonymously with "nature conservation". It stands for a whole cluster of concepts, some old ones and some that are denoted by neologisms, derived as specifications or qualifiers from "biodiversity". Not the least on behalf of that, biodiversity with some right has been called a buzzword. There are hundreds of papers and whole books that try to sort out what biodiversity is (and what it is *not*), how it should be assessed and valued (e.g. Gaston and Spicer 2004, MacLaurin and Sterelny 2008, Maier 2013).

The biodiversity cluster is therefore a very broad one that encompasses a great variety of **objects** – or more broadly phenomena – from specific units (especially species, but also habitats and ecosystems as well as genes; see CBD 1992[10]) on the one hand to "diversity as such", i.e. the variety of these units (most prominent also here: species richness) and the property of being different (Solbrig 1991, p. 17). But it is not only this diversity that is in the focus of biodiversity and related terms but also the uniqueness of each species (or gene, ecosystem, etc.), which is seen as a reason to protect them. In addition, biodiversity also includes functional diversity (functional types of organisms and their traits; e.g. Cadotte et al. 2011). And, in a broad sense, biodiversity is equated simply with "(living) nature", as Takacs (1996) has elaborated, or with "life on Earth" (cf. in the iconic MA scheme; MA, 2005, p. vi; see Figure 3.4). Also single species or organisms are perceived as symbols of biodiversity or as biodiversity, e.g. natural monuments like old trees. It is obvious that there can be no single numeric measure nor indicator to describe these various conceptualisations of biodiversity and the objects they refer to, also not for some of its more specific expressions such as functional diversity or ecosystem diversity. Depending on the specific interpretation of biodiversity, the pragmatic objects thus vary, as do the ultimate objects: often the species or (bio)diversity "as such", but also ecosystems and their functioning or – not so rarely – human well-being.

This variety is also reflected in the **values** embraced. Far more than being a merely scientific concept, "biodiversity" is intimately linked to values. It was already created in 1986 (see Takacs 1996, p. 37ff) with an emphasis on saving species and/or living nature: it was thus value-laden from the beginning. However, as already some qualifiers of biodiversity (especially functional diversity) show, biodiversity is not linked to one specific type of value but to an array of different values (e.g. Norton 1987, Maier 2013; see also the preamble of the CBD), instrumental as well as non-instrumental ones. For many conservationists, biodiversity (especially species diversity) is considered to have an intrinsic value. However, when it comes to functional diversity and relating biodiversity to ecosystem services, it is often implicitly assumed that "high" biodiversity is generally "good", not the least for human well-being, and is certainly better than low biodiversity (Maier 2013, p. 132f). In other cases, valued "biodiversity" refers to "typical" biodiversity (typical for specific places and/or ecosystems), which might well be low for some systems or regions (Jax and Heink 2015). "Typical" biodiversity often links to relational values: the experience of traditional ecosystems, of "home", as a cultural expression. This is at least one background of the sharp distinction often made between "native" and "exotic" biodiversity, with native diversity valued positively and exotic diversity negatively (see Box 4.1). However, for other authors, for whom the native–alien dichotomy seems irrelevant, the same alien species are a positively valued part of biodiversity, either simply because they are unique entities in evolutionary terms or because they contribute to the functional diversity of a system (e.g. Schlaepfer et al. 2011). Functional diversity also contributes to the instrumental values of biodiversity.

Like the ecosystem services concept, biodiversity is thus a hybrid concept oscillating in its use between a descriptive scientific and a value-laden character, benefiting from both the respectability provided by the former and from the motivational drive derived from being a conservation target ("biodiversity is good") (Eser, 2002). This hybrid character was and is consciously used in a **strategic** manner, not the least – in policy and the broader public – to overcome prejudices against "nature conservation" as a static and inflexible approach excluding human uses. Edward O. Wilson, in the preface of the book *Biodiversity* (Wilson 1988 – being the proceedings of the 1986 conference for which the term "biodiversity" was coined; see Section 5.1), even praises biodiversity for the link between conservation, politics, and economy, postulating "a new alliance between scientific, governmental, and commercial forces – one that can be expected to reshape the international conservation movement for decades to come" (Wilson 1988, p. iv).

The background for introducing and pushing biodiversity as a strategic concept was certainly – for most of its protagonists (see Takacs 1996) – a general concern about the accelerating loss of species. This concern – already present at the onset of Western conservation and valid to this day – is very often motivated by a deep feeling of wonder and of responsibility for species and nature as a whole (McCord 2012). The arguments brought forward for protecting species and other natural variety, however, are diverse, as already visible in the list of arguments given in the Ehrlichs' book (1981, p. 6), also including commercial ones and those related to ecosystem services (meant there as a precaution to avoid the collapse of ecosystems on which humans are dependent).

In consequence, there is also not one idea and **narrative** of human–nature relationship that can be attributed to biodiversity and the whole cluster. I named "biodiversity" as one of the paradigmatic concepts of the major narrative of a "Caring for other living beings (and natural features) is part of a good human life". This is correct in many respects, but certain understandings of concepts in this cluster also refer to other human–nature relationship as described in other major narratives: to "Safeguarding necessary and useful nature", when biodiversity is seen as a resource; to "Steering nature in a responsible way", when (functional) biodiversity is seen as the basis for the sound functioning of ecosystems; for "Being at home in nature", when it comes to valuing "typical" biodiversity.

Cluster "Wilderness"

Cluster: wilderness, wild, wildness, rewilding, pristine, untrammelled, Half Earth, *Prozessschutz* [process conservation], wilderness experience

The concepts of wilderness and wildness as such have a long history, going back to antiquity. They are not scientific but cultural concepts. Only in the 17th century wilderness became positively valued (Kirchhoff and Vicenzotti 2014). Before that, wilderness and wild land was mostly something to avoid or to domesticate. In spite of predecessors such as Henry David Thoreau, for whom wilderness, and even more wildness, was an important topic, its use in conservation originates in late-19th-century North America, especially with John Muir. Much has been written about the history of the wilderness idea (Nash 1967/2014, Oelschlaeger 1991, Haila 1997, Lewis 2007). Seen on the one hand as a major landmark and an ideal for conservation, it is at the same time highly contested for excluding people and promoting unrealistic romantic ideas as conservation goals, especially when applied in developing countries (Guha 1989).

Different understandings exist of what characterises wilderness. Overall, as shared by all authors, wilderness refers to the state of an area with no or only minimal human influence. Beyond that, differences start. Does an area, in order to be called wilderness – or even to be designated as such – have to be *pristine*, in the sense of being untouched by humans, at any time? Or is it sufficient that only "modern" humans did and do not affect it? As mentioned before, in the early 19th century George Catlin and others wrote about an "Indian wilderness" *including* native people. Can something also be or become wilderness when it was inhabited before but when it is free of human influence only from now on? If yes, which minimum standards for the area are necessary at the time it is designated as "wilderness"? Beyond "pristine", other words to characterise wilderness are "primeval" and "natural", both terms with various different interpretations (see Section 4.1), and also "untrammeled".

According to Scott (2000/2001), first attempts to come to more precise definitions of wilderness – for the purpose of management – started in the 1920s. The decisive formal guideline to which many conservationists refer to today, however, is the US Wilderness Act from 1964. The definition of wilderness given there is helpful for elaborating various nuances of meaning of the concept. So it's worth quoting it. According to section 2c of the Wilderness Act, "wilderness" is defined in the following way:

> A wilderness, in contrast with those areas where man and his works dominate the landscape, is hereby recognized as an area where the earth and its community

of life are untrammeled by man, where man himself is a visitor who does not remain. An area of wilderness is further defined to mean in this Act an area of undeveloped Federal land retaining its primeval character and influence, without permanent improvements or human habitation, which is protected and managed so as to preserve its natural conditions and which (1) generally appears to have been affected primarily by the forces of nature, with the imprint of man's work substantially unnoticeable; (2) has outstanding opportunities for solitude or a primitive and unconfined type of recreation; (3) has at least five thousand acres of land or is of sufficient size as to make practicable its preservation and use in an unimpaired condition; and (4) may also contain ecological, geological, or other features of scientific, educational, scenic, or historical value.

The history of the Act and in particular of this passage are well-documented (e.g. Scott 2001/2002, Cole 2000). The wording of the Act has been drafted with much care and clear intentions. According to this, the first sentence constitutes the "ideal" or the "essence" of the wilderness concept, whereas the second sentence is "descriptive of the areas to which this definition applies" (Scott 2001/2002, p. 75, citing Howard Zahneiser, the author of the Act) and is less demanding.

It is obvious from both sentences that wilderness is exclusive of humans, "man" being a "visitor who does not remain", and "without human habitation". "Untrammeled" (not the most common word in the English language) was supposed as "not being subjected to human controls and manipulations that hamper the free play of natural forces" (ibid.). As many wilderness protagonists are at pains to emphasise, it does *not* refer to a pristine state, "pristine" being a term that is found nowhere in the Wilderness Act. Less clearly defined are the terms "primeval character" and "natural conditions", which the second sentence refers to. Cole (2000, p. 79) interprets and distinguishes the different terms used in the Act as follows:

> I consider wilderness ecosystems to be "wild" and "untrammeled" to the extent that they are free from intentional human manipulation and control; "natural" to the extent that post-aboriginal human influence is lacking; and "primeval" to the extent that their present state is consistent with historic pre-aboriginal conditions.

According to this view, the "pre-aboriginal state" is still the ideal, but the actual area must only be consistent with this. Especially the question of what is natural, however, is highly controversial in conservation (see Section 4.1) and one of greatest sources of misunderstanding between conservationists.

What is also mentioned in the definition of the Wilderness Act is large size as a criterion for wilderness areas ("has at least five thousand acres of land or is of sufficient size as to make practicable its preservation and use in an unimpaired condition"). This one of the things that distinguish it from *wildness*. Finally, another feature that pops up in the Act is that wilderness should normally not be continuously managed ("without permanent improvements"), sometimes also expressed in scientific parlance as being "self-sustained" or "self-regulating", which is at the same time seen as one characteristic of its "intactness" or "integrity" (together with the full complement of native species).

While wilderness is usually related to a specific area, *wildness* is not. "Wildness" is more a characteristic of something, as is also "wild" (which might also be a behaviour). Wildness can also be letting ecological processes running free, and it can be a characteristic of individual animals (in contrast to domesticated one; see case study in Section 3.1). An area exhibiting wildness can thus be a brownfield within a city, an unguided succession on a former mining area, even a wild garden. Wildness may also be experienced in novel ecosystems (see later). Especially when wilderness is equated with a pristine state of an area, this makes a decisive difference. Every area of wilderness in this sense is certainly also wild, but not every wild area is wilderness. Thoreau, for instance, in his famous dictum "In wildness is the preservation of the world" (Thoreau 1862/1992, p. 30) spoke of wildness, not wilderness.

Related to these concepts is that of *rewilding*. I have been struggling if rewilding should better be subsumed to this cluster or to the "restoration" cluster. Eventually, I listed it in both clusters but will deal with it here in some more detail. Even though rewilding is indeed very often seen as a kind of restoration – which the "re-" indicates –, it is useful to discuss it here, because the distinctions made earlier, between wilderness and wildness, help to grasp some of the differences in meanings attributed to the term "rewilding".

The concept of "rewilding" already originated in the 1980s but has won increasing attention during the last years. As with the other restoration concepts its point of departure are "damaged" ecosystems. The different directions and definitions of rewilding strongly reflect the differing historical and ecological contexts of North America on the one hand and Europe on the other (for overviews on the spectrum of definitions, see Jørgensen 2015, Lorimer et al. 2015). In North America, rewilding in fact starts from *wilderness* areas, in which, however, some major species (keystone species) were seen as missing, in particular, large predators like wolves and bears (Soulé and Noss 1998). These species should be reintroduced to restore the original wilderness ecosystems, by re-establishing trophic structures that were lost through human actions ("trophic rewilding"). In contrast, many European projects typically try to bring more *wildness* to the area, e.g. by passive restoration of abandoned agricultural areas (Navarro and Peirera 2012). In these areas, natural processes should be left running without human interference, sometimes aided by an initial introduction of large herbivores. These areas are mostly much smaller than American wilderness areas, and are very often embedded in cultural landscapes (see case study in Section 3.1).

The **values** connected to wilderness and wildness are diverse. Although often especially the intrinsic value of wilderness (or the species therein and in wild places) is emphasised, in fact a very broad spectrum of values has been connected to wilderness and wildness and is used to justify its protection. Nelson (1998) lists 30 arguments for protecting wilderness, which cover the whole spectrum from utilitarian via relational to intrinsic values. Experiential and educational values were strong reasons for embracing wilderness and wildness from the beginning: to experience untamed, uncontrolled nature and to learn from it. Strictly instrumental values, however, such as the provision of some ecosystem services, are more the exception (but see Pettorelli et al. 2018) or more a strategically motivated add-on to relational and intrinsic values perceived in wildness and wilderness.

This diversity of values makes it difficult to identify what the "real" **objects** of wilderness and wildness protection are. The pragmatic objects are certainly the areas to be safeguarded as wild and sometimes these are also the ends, when wilderness and wildness are valued for their own sake. More often, conservationists protect wilderness (areas) for the sake of species, or more general (native) biodiversity, these being the ultimate objects (but see Sarkar 1999 on possible trade-offs between wilderness conservation and biodiversity conservation). This seems to be the case, e.g., for Edward O. Wilson's (2016) notion of "Half Earth", which demands to protect half of the Earth's surface; at least for Wilson himself, wilderness protection appears to be instrumental to species protection. Likewise, wilderness is often instrumental for human needs, be they spiritual or recreational, thus making humans and human well-being the ultimate object at which wilderness protection aims. Some authors even see the (natural) processes and not so much (specific) species as the major objects and focus of wilderness protection; this has also been the idea of *Prozessschutz* (process conservation) in Germany (Scherzinger 1990; but see Soulé 1996 for a critical stance regarding a primacy of protecting processes).

In contrast to, e.g., the ecosystem services concept, wilderness and wildness have almost never been used in a mainly **strategical** manner. As briefly mentioned earlier, it is only the other way round that different benefits of wilderness, such as ecosystem services, have been used to justify rewilding or the protection of the wild.

The absolutely dominant **narrative** of human–nature relationships that underlies wilderness concepts is clearly that of "Humans apart from nature", often jointly with that of "Caring for other living beings is part of a good human life". "Caring" in this case is understood as a distant relation, with a minimum of physical interactions with nature. Only very rarely are others of the above narratives involved. When it comes to wildness, however, the emphasis is sometimes much less on "Humans apart from nature" but may also link to "Gardening nature in novel ways".

Criticism has been raised against the wilderness concept on behalf of several arguments (see, e.g. Callicott 2000b, Wuerthner et al. 2014). A main concern with the wilderness idea is that it excludes humans; additionally there are allegations that it follows a naive ideal of a pristine, untouched nature. While many wilderness protagonists today reject the latter allegation, the practice of wilderness conservation in fact has provided reasons for concern. Especially in the early days of conservation (see Section 2.2), indigenous, but also "modern", people have been removed from national parks and other areas, in order to arrive at the ideal of uninhabited wilderness (Spence 1999, Reich 2001). It still is a problem today (see Dowie 2009). Concerns about removing people from wilderness areas and other protected areas have, not the least, been raised against the Half Earth movement (Wilson 2016), which demands to protect half of the Earth's surface – even though it is not clear under what degree of protection (see Büscher et al. 2017 for a critique). Certainly, setting aside areas as wilderness is not a bad thing per se and one should not throw out the baby with the bathwater here. Referring more to *wildness* than to wilderness may help to overcome this problem, as wildness does not exclude humans as much as wilderness (Haila 1997, Jørgensen 2015).

Cluster "Cultural landscapes"

Cluster: cultural landscapes, traditional landscapes, homeland and *Heimat*, biocultural conservation, belonging, environmental identity, sense of place, semi-natural areas, natural and cultural heritage, [agroecosystem]

BOX 3.7 THE LANDSCAPE CONCEPT: DIFFERENT MEANINGS AND HISTORICAL ORIGINS

"Landscape" is an old term. It is used with various meanings. I will, following Berr and Schenk (2019), focus here only on three major ones (for more detail on the history of the landscape concept see Olwig 1996, Trepl 2012). The word originated at least in the Middle Ages in the Germanic languages as "*lantschaft*", "*lantscaf*", or "*landschap*", later becoming "*Landschaft*" in modern German and "landscape" in modern English.

Originally, landscape had a *territorial and law-related meaning*, designating some settled area with common social conditions, and institutions (especially jurisdiction). A second meaning came up during the early modern area, focusing on *landscape as an aesthetic category*. First designating a painting of natural scenery, it soon became transferred to the object of the painting itself. By means of this artistic perception of nature – not only in painting but later also in poetry – landscape received a strong emotional and symbolic dimension; landscape was even seen as a mirror of the soul, especially from the 18th century on. It was in this context that ideas of a close unity between people and the land they inhabited were developed, as shaping personal and social identity. This was manifested in the early protection of cultural landscapes (see Section 2.2). Only by more and more removing the aesthetic, emotional (and even social) characteristic of the landscape idea, a third meaning of landscape could develop, in geography. Here, landscape was seen as the "totality" of an area, which may be *objectively assessed as an empirical object*. This is also the

> way it partly found its way into ecology. In a well-known American textbook of "landscape ecology", landscape is defined as "a heterogeneous land area composed of a cluster of interacting ecosystems that are repeated in similar form throughout" (Forman and Gordon 1986, p. 594). In contrast, European textbooks on landscape ecology still see the aesthetic and symbolic dimensions of landscapes as essential characteristics for describing and understanding them (see Kirchhoff et al. 2013).

The concept of *cultural landscapes*, as landscapes shaped jointly by humans and non-human nature, is one that was developed outside of and previous to a conservation context. It was first used by the German geographer Carl Ritter in the early 19th century, the latest in 1832, as *Culturlandschaft* (*Kulturlandschaft* in modern German orthography) (Kirchhoff et al. 2013, Potthoff 2013). The American geographer Carl Sauer in 1925 transferred the concept as "cultural landscape" into English-speaking geography. As described in Section 2.2, the idea of cultural landscape played an important role in early European conservation and does so to this day (Figure 3.5). In Europe, it is also a central concept in landscape planning and management. In contrast, it does not have much popularity in the mainstream of American conservation and conservation biology. Internationally, however, the notion "cultural landscape" and related concepts find increasing attention in conservation (Jacques 1995). It was promoted at first most prominently via UNESCO's Cultural and Natural Heritage Convention (1972) and its follow-up processes, but increasingly convergences between this convention and IUCN's protected areas policies have developed (Phillips 1998, Taylor and Lennon 2011). The term "cultural landscape" has been used with different meanings. Kirchhoff et al. (2012, p. 53) distinguish four major meanings (see also Jones 2003):

(1) every landscape that has been (perceptibly) modified by human cultural activity; (2) landscapes shaped by traditional forms of land use that are valued as putatively representing the result of a harmonious and unique human-nature relationship […]; (3) landscapes designed and created intentionally by humans, such as parkland landscapes constructed for aesthetic reasons […](4) landscapes that are cultural just because aesthetic, cultural or spiritual meanings have been attached to them, irrespective of whether they have been (perceptibly) modified by human activity.

As Kirchhoff and colleagues state, meanings 2–4 come close to the types of cultural landscape defined since 1992 in the implementation guidelines of UNESCOs World Heritage Convention (UNESCO 2021, pp. 22–23):

- Meaning 3 corresponds to UNESCO's category (i): "landscape designed and created intentionally by people"

116 Analysing conservation concepts

- Meaning 2 corresponds to UNESCO's category (ii): "organically evolved landscape"
- Meaning 4 corresponds to UNESCO's category (iii): "associative cultural landscape".

UNESCO further divides category (ii) into "relict landscapes" and "continuing landscapes", the latter defined as a landscape

> which retains an active social role in contemporary society closely associated with the traditional way of life, and in which the evolutionary process is still in progress. At the same time it exhibits significant material evidence of its evolution over time.
>
> *(ibid., p. 23)*

While UNESCO's World Heritage Convention is focusing on landscapes of "outstanding universal value" (UNESCO 2021, p. 22), cultural landscapes are often also "everyday landscapes" (Taylor and Lennon 2012) that have long been taken for granted but are now often endangered. The question is, however, which of these four types of cultural landscapes are also landscapes at which nature conservation aims or should aim. Meaning 1 of Kirchhoff et al. (2013) is so broad that some authors have argued that it is not necessary to add the term "cultural" to it, as practically

FIGURE 3.5 Traditional cultural landscape in southern France (Parc Naturel Régional des Alpilles, near Mouries). The landscape is a mosaic of different habitats, some parts actively managed, others not. Photo: Kurt Jax.

every landscape is influenced by humans (see discussion in Jones 2003). In fact this meaning would also include highly industrialised agricultural landscapes, which certainly (as *landscapes*) are beyond the scope of nature conservation. An explicit emphasis on the "cultural" is necessary because especially in American landscape ecology "landscape" is often defined purely ecologically without any mentioning of the social or cultural – very much in contrast to the European tradition (see Box 3.7). Kirchhoff et al.'s type 2 ("landscapes shaped by traditional forms of land use that are valued as putatively representing the result of a harmonious and unique human–nature relationship.") is the type that is closest to the common meaning of cultural landscape in nature conservation.

In cultural landscapes, the mutual interactions of humans and their surroundings mean that they shape each other in both directions: the biophysical landscape partly shapes human culture and society, and in turn human society and humans, with their culture and actions, give new form to biophysical nature. Part of this mutual interactions gives rise to place-based identity and belonging (see, e.g. Taylor and Lennon 2012). In contrast to a purely systemic approach, cultural landscapes are not just places where resources (such as food) are taken from nature but include major symbolic dimensions. These symbolic dimensions are not "services" because they are not "provided" by non-human nature but lie in the very relation between humans and the non-human surroundings they interact with. Cultural landscapes also are the places where the *memories* of these interactions are located (see the title of Simon Schama's famous 1995 book *Landscape and Memory*). For that reason, a specific cultural landscape is not just a reproducible *system*; it has *history* and leads to a "*sense of local distinctiveness*" (Taylor 2008). This constitutes an important difference to the concept of social-ecological systems (see later). It, however, also raises questions such as how old a landscape must be to count as "traditional" or "organically evolved".

At first sight, the **objects** of "cultural landscapes" and related concepts are the landscapes themselves. The landscape is the pragmatic object, understood, however, not just as a spatial pattern (including cultural objects) to be preserved but also as the intricate ongoing interactions between humans and nature. The ultimate object to protect varies, however. It might be human well-being or a good human life (in a broad sense, including good human relations with nature), it might also be specific species, when cultural landscapes are used "instrumentally" to protect endangered species. Thus, in many conservation areas (*Naturschutzgebiete*) in Germany, grassland sites that had a long history of human management, and which would normally reforest without human interference, are kept open by continuing or mimicking traditional land use methods. The purpose is here one of aesthetics and local identity but to a considerable degree also to prevent the (local) extinction of species, e.g. rare orchids. Many of these species became abundant as a by-product of extensive historical agricultural practices. They now are seen as an important biological – and cultural – heritage to be preserved. Protecting cultural landscapes can thus also be a kind of tool for species protection or even for maintaining a sustainable use of ecosystem services.

In terms of **values**, relational values are the most conspicuous ones for the concepts in this cluster, as the interactions between humans and non-human nature, with all the symbolic values they create, are also an end in itself here. That does not preclude that also use values play an important role, e.g. from agricultural products, but they do not dominate the notion of cultural landscapes *as conservation objects*.

The dominant human–nature relationship **narrative** here is that of "Being at home in nature". Depending on the context, also others, such as "Safeguarding necessary and useful nature" and "Caring for other living beings..." can be important. Cultural landscapes may even be a concept that can integrate several of these narratives. The other way around, narratives about specific places can play an important role in the constitution of the cultural landscapes as they carry on history and memories in the form of specific stories.

The term "cultural landscape" and its formalisation as a (cluster of) concepts for conservation is of Western origin. The ideas and the practices to which this concept refers, however, have long been present all over the world, from traditional low-impact agriculture in Europe through rice terraces in the Philippine Cordillera to the aboriginal cultural landscapes of Australia (see various examples in von Droste et al. 1995, Taylor and Lennon 2012; see also Section 4.3). Even though the conservation of cultural landscapes has been a prominent theme in many countries, some conservationists might still argue that conserving cultural landscapes is not *nature* conservation at all. This is to a high degree a matter of what is considered as nature and what is natural, an issue to which I will come back in detail in Chapter 4.

Cultural landscapes have sometimes been decried as static and their protection as "museum-like", perpetuating or even mimicking "traditional use" as a kind of management method (as in the example of the protected German grasslands described above). As UNESCO's subtype of "continuing landscapes" and many active examples show, this is not necessarily the case. New modes of interactions may perpetuate a long history of such human–nature relationship without sharp breaks. However, difficult questions remain: do wind turbines belong to a cultural landscape, as traditional windmills or even castles (often medieval *military* facilities!) do now? Maybe not today, but may be in 50 years they may be judged as a characteristic feature of such landscapes, even symbolising a new sustainable relation of humans with nature – or to the contrary – they may be seen as a symbol for sacrificing animal habitats (bird habitat) at the cost of undiminished human energy demands. Much will depend here both on a prudent and ecologically informed planning and on how local people are involved in this.

A concept that, depending on its specific interpretation, stands half-way between the cluster of cultural landscapes and that of ecosystem management, or even may be part of both clusters, is that of "biocultural conservation" (see Bridgewater and Rotherham 2019 for the history of the concept and its definitions). It focuses on "bio-cultural landscapes" (Bridgewater and Rotherham 2019) or "biocultural co-inhabitation" (Rozzi 2013, 2015b) of areas, and it emphasises the interplay between cultural and biological diversity, especially in traditional land use systems.

An understanding of "biocultural approaches" more aligned to the idea of social-ecological systems (see cluster "ecosystem management") has been described by Gavin et al. (2015).

Cluster "Ecosystem management"

Cluster: ecosystem management, ecosystem-based management, ecosystem approach, ecological integrity, ecosystem health, ecosystem integrity, ecosystem degradation, social-ecological systems, socio-ecological systems, biocultural conservation

An important characteristic of the concepts in this cluster is the explicit focus on whole ecosystems or even social-ecological systems, be they wild and "natural" or not. Ecosystem management, at least in its current form and under this term, is one of the newcomers in conservation. Like the other terms that denote the major clusters, the word stands for a whole programme of how to deal with nature, and by some authors (e.g. Stanley 1995, Lackey 1998) has even been called a "paradigm". The concept has a strong science-based background. However, as many authors have emphasised (e.g. Grumbine 1994), it is not only about science but also about humans as part of the system and about societal values, choices, decision processes, and institutions.

The idea of ecosystem management was at first developed as a purely ecology-based approach. It started from the insight that even large national parks like Yellowstone were not big enough for protecting large, wide-ranging mammal species such as the Grizzly bear (Craighead 1979). The first important and at the time revolutionary shift going along with this insight was to move from managing specific selected species or natural features towards managing whole ecosystems. This involved an emphasis on ecological (i.e. "natural") boundaries for management areas (e.g. determined by the home ranges of large mammal populations) instead of administrative ones. This, in turn, necessitated interagency cooperation in management (e.g. between the US Forest Service and the National Park Service), which soon became a second hallmark of ecosystem management. A further decisive element of ecosystem management today is the inclusion of stakeholders and of different knowledge forms beyond scientific knowledge (e.g. traditional and local knowledge) in planning and implementation. From the late 1980s on, a great number of concepts and projects on ecosystem management and, as another term, ecosystem-based management were developed and applied, especially in the USA and Canada. In the 1990s, ecosystem management also reached the political arena both in some national policies (e.g. the establishment of the Federal Ecosystem Management Initiative in the USA), as well as in international contexts. On the international level, so-called "Ecosystem approaches" have been codified as a preferred mode of management by several international programmes, such as those of IUCN, UNEP, UNESCO's Man and the Biosphere Programme, and the CBD, in the latter in the form of its Malawi principles and the related "Ecosystem

Approach of the CBD" (see UNEP/CBD 2000). In this whole process, the meaning of "ecosystem" more and more changed from ecosystem as a *scientific object* towards ecosystem as *management perspective*. Overviews on the philosophy, methodology, and practice of ecosystem management can be found in Grumbine (1994), Boyce and Haney (1997), Lackey (1998) and – for a more recent review of its current challenges – DeFries and Nagendra (2017).

As a related concept, crucially emphasising the inclusion of humans and their societies into a common management unit, the concept of socio-ecological or social-ecological systems was developed. Here, not only are humans seen as part of the ecosystem but their whole societal processes (including economic and governance processes) are dealt with as subsystems of the larger system to be managed. The concept goes back to Nobel laureate Elionar Ostrom and her work on the governance of commons (especially natural resources), institutions, and collective action in the 1980s and 1990s, which required a systematic connection between ecological and social variables, as parts of one common system (Partelow 2018). Other researchers approached such linked social and ecological systems from a resilience perspective (Berkes and Folke 1998, Colding and Barthel 2019). The research communities of ecosystem (based) management and that of social-ecological systems are, however, rather distinct from each other, even though what is meant by "ecosystem" in most ecosystem-based management approaches are in fact social-ecological systems. There have thus been some debates on the difference of the concepts ecosystem management, ecosystem-based management, and ecosystem approach, as well as the ideas of managing socio-ecological or social-ecological systems. The latter listing in fact constitutes a gradient in terms of an increasing degree to which individual and societal human activities are seen as part of the system, already emphasised in the terms used. The general approach, however, is similar in all these cases and follows the characteristics described.

One difference between ecosystem (based) management and social-ecological systems relates to the *supporting concepts* by which the desired proper functioning of the systems in focus is described. Ecosystem (based) management approaches commonly refer to ecological or ecosystem health, or to ecosystem integrity. These concepts are rarely used in the context of managing social-ecological systems, which instead refer to the resilience and sustainability of the systems as the target to be maintained. Also, social-ecological systems approaches tend to be less focused on biodiversity and more on resources (Rissman and Gillon 2017). All these supporting concepts also have a normative dimension (see Jax 2010). They are difficult to operationalise, often used without much qualification (sometimes as mere buzzwords) and in fact require societal choices for providing management targets, something which, e.g., the first of the Malawi principles emphasises by stating: "The objectives of management of land, water and living resources are a matter of societal choice" (UNEP/CBD 2000).

The same things (the necessity of societal choices) holds also for the **objects** of the approaches discussed under this cluster. As stated already, e.g., by Lackey

(1998) and explained in much detail in Jax (2010), "the" ecosystems or "the" social-ecological systems are not given by nature as such but require decisions about what belongs to the system and what not, how the system is bounded in space, etc. What exactly the respective object is, is dependent – not only but to a considerable degree – on the question(s) one asks and the problem one wants to tackle. An often unanswered question is: are mainly processes (Callicott 2000a) or species in the focus of management? Thus even the pragmatic object is not clear, much less its ultimate object. Is it human well-being, is it biodiversity (in whatever definition), is it in fact the "whole system" – or all of these; and if the latter is the case, how to strike the balance? Even when focusing on protecting biodiversity by means of ecosystem management (as common especially in the early days of ecosystem management), is the target *all* of biodiversity or only "native" biodiversity (Grumbine 1994 and others postulated the latter)?

Almost obviously, given the variety of ultimate objects also the **values** embraced by proponents of the ecosystem management approaches differ from utilitarian to intrinsic. Stanley (1995) and also Lackey (1998) even spoke about a dichotomy between anthropocentric and biocentric approaches of ecosystem management.

Nevertheless, the potential dichotomy in values also hints to different **narratives** of human–nature relationships found in different interpretations of the concepts in this cluster. Many are related to the "Steering nature in a responsible way" narrative, but some uses of ecosystem management and social-ecological systems concepts are clearly extensions of a more sophisticated resource management and thus would be linked to the "Safeguarding necessary and useful nature" narrative. Others might even be seen as expressions of yet other human–nature relationships (narratives), obviously with the exception of the "Humans apart from nature" narrative. In the concepts of this cluster, humans are clearly considered as a part of the *system*, and so is nature perceived. But the desired role of humans differs substantially between various expressions of ecosystem management or social-ecological systems. Are humans the master-head with the right (or even duty) to steer nature, especially in the face of the Anthropocene, or are humans humble members of nature who steer it in the interest of all life (see also Box 3.6 on stewardship concepts)? Aldo Leopold was credited by some authors as an important forerunner of both of ecosystem management (Callicott 2000a) as well as of the notion of social-ecological systems (Meine 2020). Leopold did not use the term ecosystem but he referred to the "community" and the "land", both of them meant as including humans. In fact, he anticipated several of the ideas discussed in this section. For Leopold, this all was connected to a new ethics and a new role of humans:

> In short, a land ethic changes the role of *Homo sapiens* from conqueror of the land-community to plain member and citizen of it. It implies respect for his fellow-members, and also for the community as such.
>
> *(Leopold 1949, p. 204)*

Leopold's ideas might thus also bridge several of the narratives described earlier.

Cluster "Novel ecosystems"

Cluster: novel ecosystems, ecological novelty, designed/designer ecosystems, no-analogue systems, emerging ecosystems, gardening metaphor, hybrid ecosystems

Concepts in the cluster "novel ecosystems" are a very recent addition to programmatic concepts for conservation. Notwithstanding some earlier authors using the term, the concept was formalised and brought into the conservation discussion mainly by Richard Hobbs and colleagues (Hobbs et al. 2009, 2013a, Lindenmayer et al. 2008). Acknowledging strong and persistent human-induced changes in nature, the novel ecosystem concept was developed as an approach for the management of transformed ecosystems and landscapes, in those places where classical restoration was not possible any more, or only with a very high effort. Changes in abiotic conditions, e.g. by mining activities, or biotic conditions, such as the spread of non-native species, are examples for drivers that after some time created "novel" or "no-analogue" ecosystems, i.e. ecosystems that had not existed in the same combination of species and abiotic conditions in the past. These ecosystems have usually been considered as being of no conservation value, as "degraded" or even as "trash ecosystems". Protagonists of the novel ecosystem concept, however, are aiming at acknowledging that also these systems can be of some conservation value. Although definitions vary largely (see, e.g. table in Truitt et al. 2015), the following criteria are mostly named for something to be considered a "novel ecosystem" (e.g. Higgs 2017):

1 As a result of human impacts, there is a new combination of species (often including exotic ones) and/or abiotic factors that did not exist in previous times.
2 The system must be persistent (or "self-sustaining" or "self-reinforcing"), i.e. perpetuate itself without any ongoing human management.
3 It cannot be reversed to a previous, historical state.

There are considerable differences in the understanding of "novel ecosystems" with respect to each of the above criteria. In fact, each of the supportive concepts used to define novel ecosystems (e.g. human impact, self-sustaining etc.) requires some clarification on their own (see Morse et al. 2014 for a good discussion). Some scientists (e.g. Morse et al. 2014), for reason of clarity and practical usefulness, restrict the *human impacts* by which a novel ecosystem can be created to *direct* impacts (no matter if intentional or not). They explicitly exclude indirect impacts such as human-induced climate change or nitrogen deposition from global air pollution; others do not make this distinction (e.g. Fox 2007). There is also discussion what *"irreversibility"* means, how to ascertain it, and if it is only related to ecological irreversibility (such as major species driven to extinction, or completely altered soils) or also to social and economic irreversibility (socially undesirable and not enforceable, or at unbearable economic costs). Connected with this is the question as to whether there are real *thresholds* towards an irreversible state of the ecosystem, or if the degree of human impact constitutes more of a gradient (Miller and Bestelmeyer 2016). This is also reflected in the idea of "hybrid ecosystems", as systems that contain some novel elements (e.g. exotic species) as well a historical ones, and which may still be restored to a historical system (e.g. Hobbs et al. 2014). Many authors, however, consider the term "hybrid ecosystems" to be too fuzzy and simply speak of "impacted ecosystems" (e.g. Morse et al. 2014; see Figure 3.6). As part of this cluster, then also the notion of "designed (or designer) ecosystems" has been discussed. Such systems are in general distinguished from novel or impacted ecosystems by the fact that they were created intentionally for human purposes and are maintained actively by human management. Examples would be urban parks, agricultural systems, or artificial lakes. Both impacted as well as designed ecosystems can develop into novel ecosystems if human agency ceases (dashed lines in Figure 3.6). The terms "no-analogue" or "emerging" ecosystems are used either as synonyms for novel ecosystems, or sometimes for describing "new" ecosystems resulting from climate change (instead of by direct human intervention). Issues of

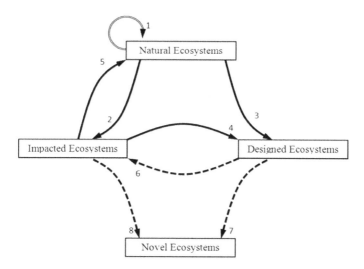

FIGURE 3.6 A suggestion for distinguishing novel ecosystems and related concepts, focusing on the role of human agency in the creation and maintenance of altered ecosystems. Arrows: (1) natural ecosystem processes, (2) unintentional alteration or intentional degradation, (3) altering ecosystem services for clear human gains, (4) environmental remediation, (5) ecosystem restoration, (6) lack of maintenance, and (7 and 8) novel succession.

Figure reprinted from Morse et al. (2014), via Ecology and Society, Creative Commons Attribution 4.0 International License.

clearly delimiting the different concepts within this cluster – both conceptually as in practice – prevail, however.

The pragmatic and direct **object** of conservation and management of novel ecosystems is clearly the ecosystem (however delineated), but the ultimate objects and the values which are to be protected vary. The objects are partly seen to be ecosystem services provided by the novel systems but also (endangered) species. Higgs (2017, p. 10), e.g., emphasises that "the ecosystem comes first", meaning for him that for "novel ecosystems the focus is similar to ecological integrity and biodiversity goals for ecological restoration" (ibid.) and not on ecosystem services and immediate human interests. Hobbs et al. (2009) see both "biodiversity" as well as ecosystem services as potential goals of managing novel ecosystems (see also Evers et al. 2018). The **values** attributed to novel ecosystems can thus be both instrumental as wells as more to the intrinsic side of the spectrum (see the quote from Higgs 2017, above). Over time, novel ecosystem might also gain relational values. For designed ecosystems, the focus is clearly on the utilitarian side.

From the beginning, there have been controversies as to whether "novel ecosystems" is (a) a really new and useful concept and (b) if it is a *conservation* concept at all. Severe critique was raised especially from the side of restoration ecologists. They expressed doubt as to whether the concept is really something new and if the term is needed at all (e.g. Murcia et al. 2014). Beyond that, the major critique of the novel system concepts aims at its potentially negative impact on conservation, i.e. **strategic** concerns. The concern is, that by accepting "degraded" ecosystems as valuable, this concept will substitute or at least impede classical conservation and restoration efforts that aim at the original, historical ecosystems. The fear is that policymakers and the public would see no more need to put effort into maintaining or recreating "original" ecosystems. For these critics, supporting the novel ecosystem concept is a kind of defeatism. These same concerns have been raised by conservationists when the idea of restoration ecology was first promoted. Proponents of the concepts in the novel ecosystems cluster, however, have always emphasised that their intention is not to *replace* classical conservation but to broaden and supplement it (e.g. Hobbs et al. 2013b). Given that so many places on Earth are now what can be described as novel or hybrid ecosystems, especially in terms of their irreversibility to "pristine" states, these proponents state that the issue is not to *desire* such systems but to appreciate that they also can have conservation value (Standish et al. 2013).

Yet, qualms remain for some conservationists. Words matter, and care has to be taken that it is not the overly simplified idea of novel ecosystems – as being the new or even *desired* normal – is transported to the public, as an invitation to "anything goes" – in terms of blurring the difference between all types of ecosystems, including "degraded" ones (Marris et al. 2013). As Higgs (2013, p. 295) states:

> When familiar ecosystems pass over the threshold where returns to history are no longer practicable, then we need a moral basis for action. These

ideas are just emerging and will, I hope, be based on something more than "anything goes".

For example, Standish et al. (2013) emphasise that accepting the existence and value of novel ecosystems does not mean that invasive species always go unmanaged and uncontrolled.

Even though conceptual problems remain, the clear counterpoint that the novel ecosystem concept makes is that not only "natural" and "wild" systems or traditional cultural landscapes can be worth protecting for nature conservation but also new and "messy" systems that were previously unknown in that form.

The human–nature relationship **narrative** to which the terms of this cluster come closest is that of "Gardening nature in novel ways". But often, and quite explicit when it comes to designed ecosystems, also the "Safeguarding necessary and useful nature" and even the "Caring for other living beings…" narratives are related to the concepts of the novel ecosystem cluster. In contrast to the "Steering nature in a responsible way" narrative, the insight here is that one is dealing with *novel* systems, for which science may provide even less established management knowledge than for "natural" systems – an invitation for a humble attitude towards management (Marris 2015).

Cluster "Restoration"

Cluster: restoration, ecological restoration, ecosystem restoration, reclamation, rehabilitation, revitalisation, recovery, remediation, degradation, (historical) fidelity, historicity, rewilding[11]

Ecological restoration is the subject of an entire discipline, with its own international society and journals. In 2021 the United Nations have even started a UN Decade on Ecosystem Restoration (www.decadeonrestoration.org/). Depending on one's perspective, it is a part of conservation science or at least related to it (see Section 2.4).

Restoration activities as such go far back in time, e.g., with reforestation of eroding slopes, which has been done already centuries ago (see Section 2.2). The earliest effort of restoring an ecosystem *for conservation purposes* and based on ecological principles is commonly seen as the restoration of prairies in Madison Wisconsin in 1935, under the direction of Aldo Leopold (Jordan et al. 1987). But it took some more decades, until the 1980s, before restoration really was conceptualised as a major conservation-related concept, with the Society of Restoration Ecology (SER) founded in 1988 (see Hall 2005, Jordan and Lubick 2011, Martin 2022 on the history of restoration ecology and of the practice of restoration). The emphasis of restoration in the modern sense is clearly not on restoring isolated features or resources (timber, wildlife) but of whole ecosystems or landscapes.

In contrast to the other major concepts, restoration and associated concepts start from what is perceived as either destroyed or at least "degraded" ecosystems, which should be repaired. The SER defines "ecological restoration" as "the process of assisting the recovery of an ecosystem that has been degraded, damaged, or destroyed" (Gann et al. 2019, p. S7). This point is pivotal to the whole field. Restoration starts "after the fact", after severe human interventions to an "original", "natural", or "historic" ecosystem have happened. The question as to whether an ecosystem is degraded or just different (Hobbs 2016), however, is in the first place not just a scientific one but hinges on value judgements and societal choices, as does the selection of specific goals, reference states, and the desired outcomes of restoration activities (Allison 2007, Hertog and Turnhout 2018). SER provides a set of "key ecosystem attributes" to characterise (native) reference ecosystems (Gann et al. 2019). Such criteria can support judgements on reference states and the ensuing decisions for action but they still are not "given" by nature. Modern conservation thus emphasises the social dimension of restoration, e.g. by involving stakeholders in setting the goals for restoration (Perring et al. 2015, Gann et al. 2019).

An example for the existence of different possible benchmarks for restoration (and conservation) for one and the same situation can be the Juniper Heathlands in many parts of Germany. In the region where my family originally comes from, a part of the Eifel highlands, these heathlands amounted to almost 50% of the area during the early 19th century, but now they are rare and constitute an object of nature conservation measures. These heathlands, however, are the remnants of quite unsustainable management practices in the economically poor region. Forests have been destroyed since the Middle Ages, both for timber and firewood but also for and by grazing of domesticated animals (pigs, cattle, sheep). Several of the most characteristic – and protected – plants of these heathland ecosystems are in fact those which were avoided by the grazing animals. So in terms of the "original" ecosystems that prevailed in the area before human use, these are severely degraded – or in terms of the original forest, even destroyed – systems. Yet they are very often protected, both on behalf of aesthetic reasons, for protecting rare plant species thriving there, and for being a characteristic part of the traditional cultural landscape. If these areas would be left on their own, they would soon develop into a forest with a very different (and less diverse) species composition. One could say that the latter would be passive restoration towards the "original" ecosystem. But in fact, in the areas in which tree species such as pines have encroached too much (resulting in a "degraded heathland"), restoration is performed towards the traditional heathland as the reference state. As a student I participated in a campaign of restoring such places, an activity which I dubbed for myself "conservation with an axe", namely cutting out upcoming young pine trees and thus stopping succession; the whole endeavour was funded by a conservation foundation. It is, obviously, a matter of societal choices into which direction areas such as these should be restored (or not at all).

This also opens up a look on the various concepts in this cluster. I am using "restoration" as the overarching term for the whole cluster, namely for all those intentional actions aiming at bringing an ecosystem back to a desired condition, perceived as "intact", which are mostly previous, original, or "normal" conditions. Other, related or more specific terms are "recovery", "rehabilitation", "reclamation", regeneration", "ecosystem repair", "remediation", or "revitalization". They differ especially with respect to the starting point of the management (e.g. completely destroyed or only degraded), their goals (e.g. restoration of structure or process, historical conditions, or ecosystem services), but the differences between these concepts are not totally clear and agreed upon. For example, often the difference between ecosystem restoration and rehabilitation is seen in that the first refers to achieving both the structure *and* the processes or services ("functions") of the original systems while rehabilitation only recreates the processes and/or services of the reference system (King and Hobbs 2006). See also Gann et al. (2019) for the terminology of the field, and especially Higgs (2003) who provides a very good account of the conceptual difficulties of delimiting (ecological) restoration from other practices.

To specify restoration goals and to assess if restoration has been able to reach or approach them, reference systems are needed (as demanded, e.g., by Aronson et al. 1993, 1995), at least criteria for the minimal properties that a restored system should fulfil (e.g. Jordan et al. 1987, Gann et al. 2019). The restoration of ecosystems is sometimes likened to the restoration of an old painting that is degraded but whose basic features are still visible (Aronson et al. 1993). This then means the basic features of these units are targeted for maintenance or restoration. There is considerable disagreement, however, about what these basic features are. While some scientists try to restore ecosystems on a highly abstract level based on concepts such as flows of matter and energy (e.g. Brown and Lugo 1994), others consider that it is not sufficient just to restore these processes but that restoration of particular species or even local genotypes is necessary in order to capture the "essence" of the ecosystem (Ashby 1987).

The comparison of an ecosystem or a landscape to an old painting also shows the strong link of restoration to historic information. The role of such information was at first mainly that of providing detailed information for historical reference conditions (historic fidelity). Historical reference conditions are of course also important in other fields of conservation, e.g. as a goal for the management of US national parks. In a review of wildlife management in these parks, the authors around Aldo Leopold's son A. Starker Leopold, famously recommended:

> As a primary goal, we would recommend that the biotic associations within each park be maintained, *or where necessary recreated*, as nearly as possible in the condition that prevailed when the area was first visited by the white man. A national park should represent a vignette of primitive America.
> *(Leopold et al. 1963, p. 32, my emphasis)*

Historical information is and was not only used as a means to preserve or restore natural heritage but also to find reference states that might provide a benchmark for properly functioning ecosystems. Meanwhile, the use of historic information in restoration has changed and widened, e.g. by characterising some original place in terms of specific human–nature interactions, or use the information to inform scenarios of potential futures (Higgs et al. 2014).

Such conceptual change can be seen also in other respects. In the early days of restoration as a conservation concept, it was mainly focused on restoring towards "natural" ecosystems, for the "sake of nature itself" and endangered species, especially on the western side of the Atlantic. Today it spans from restoration of or towards wilderness (see Cole 2000) to the restoration for ecosystem services and natural capital (e.g. Aronson et al. 2007), or cultural landscapes (Moreira et al. 2006). Thus restoration concepts are diverse and connected to several of the other clusters discussed earlier.

Accordingly broad is also the set of objects, values, and even narratives to which restoration refers. The pragmatic **objects** are clearly ecosystems and landscapes, but the ultimate objects are either human well-being (in terms of ecosystem services but also of people's identity) or particular objects, especially (threatened) species. All types of **values** elaborated in the beginning of Chapter 3 can be motivating restoration efforts (see Clewell and Aronson 2006, Hertog and Turnhout 2018), depending on the specific approach of restoration and the specific environmental and cultural setting.

In terms of its **strategical** implications for nature conservation, the restoration concept was, especially in its early days, as contested as the novel ecosystems concept is today (where the latter, quite ironically, finds its most fervent critics among restoration ecologists). As for novel ecosystem approaches, concerns were raised that the restoration concept might compromise classical conservation efforts. Restoration might be used as an excuse for not preserving "intact" nature because, once impacted by human activities, there would still be the option to restore it (e.g. Elliot 1982). Today, given strong biophysical, cultural, as well as conceptual changes, restoration concepts are widely accepted as part of the conservation toolbox.

For strategic purposes, but also beyond that, the message of what restoration is about has been expressed by different metaphors, in particular that of designing, gardening, repairing, or healing (Hobbs and Harris 2001, Higgs 2003). The healing aspect has also been extended to societal healing or healing/restoring human–nature relationships (Blignaut and Aronson 2020) and its implementation even as an act of "public spiritual practice" (van Wieren 2008). The value-laden "healing" metaphor fits well with the goals of "healthy ecosystems" or restoring "ecosystem integrity" which, e.g., the SER guidelines emphasise as important goals of restoration.

Consequently, restoration concepts, with their departure from degraded or destroyed ecosystems and landscapes, align with the **narrative** of "Making nature whole again", but depending on the specific expression of restoration also build on and link to most of the other major narratives.

Supporting concepts

Some other, often highly complex concepts from ecology and conservation biology play important roles in specifying the major conservation concepts discussed earlier, but do not stand for or represent major "programmatic approaches" of conservation. Instead, they support these major concepts. They are relevant enough to merit discussion of their own. Examples of such concepts are those relating to *ecosystem functioning*. Ideas of (proper) ecosystem functioning are not conservation concepts on their own but are important for characterising targets and benchmarks, e.g. for concepts in the clusters wilderness, ecosystem management, novel ecosystems, and restoration. Likewise, *stewardship* (see Box 3.6) is neither a programmatic conservation concept of its own nor a narrative for one specific type of human–nature relationship, because one can find it as a desired role of humans in several concepts and narratives, in very different meanings, and reaching from animal stewardship via ecosystem stewardship, wilderness stewardship to Earth and planetary stewardship. Also, I do not consider the "*Anthropocene*" (see Box 3.4) to be a programmatic conservation concept, although a useful one for analysing different human relationship with nature, especially from the perspective of anthropology and other social sciences.

In the following, only one cluster of supporting concepts will be discussed in more detail, comprising those which I subsume under the heading of "ecosystem functioning". Other concepts are dealt with at various places, in particular those mentioned in the previous paragraph, concepts of ecological units (community, ecosystem, etc., see later and Section 4.2; landscape, Box 3.7), and of course the concepts of *nature* and *naturalness* (Section 4.1). In connection with the latter also the notions of native and exotic will be discussed (Box 4.1).

Ecosystem functioning: targets and tools in conservation

Cluster: ecosystem functioning, stability, resilience, self-regulation, self-sustaining, self-organising, ecosystem health, ecosystem integrity, planetary boundaries, ecosystem degradation

When talking about ecosystem functioning, self-sustaining ecosystems, or the resilience of ecosystems, science comes to the fore, but, in contrast to first appearance, it is not only science that comes to the fore. Ecosystem functioning, in its most generic understanding, is the overall performance, the working of an ecosystem. In this generic form it denotes any processes going on in a system composed of organisms in interaction with each other and their environment. There is thus no way in which such an ecosystem can*not* be functioning. In more narrow meanings of "ecosystem functioning", however, in which the phrase is used most commonly, it denotes *proper* functioning, i.e. *the continued performance of a system within some limits specified as reference conditions (state, dynamics, or trajectory)*. In such more narrow meanings, assessing ecosystem functioning depends on the specific ecosystem definition embraced, a specific reference condition, and specific spatial and temporal scales of observation. Ecosystems then may also be considered to be not properly functioning, or not functioning at all. There is a wide spectrum both of possibilities to define ecosystems and for defining their reference conditions, depending on specific purposes at hand (see Jax 2006, 2010). Making statements about ecosystem functioning precise and measurable in consequence involves a number of methodological and societal choices to be taken. Defining and assessing ecosystem functioning and the related terms assembled in this cluster thus involve both scientific and normative dimensions, though not necessarily moral ones. An extended treatment of the concepts in this cluster is provided in my book *Ecosystem Functioning* (Jax 2010). Only the most important conceptual issues of the concepts will be summarised here.

As to whether an ecosystem is assessed as stable or self-sustaining depends to a high degree on the definition of "ecosystem" and on the understanding what is meant, e.g., by "stable". Grimm and Wissel (1997) have analysed and summarised the multiple meanings of stability and the "Babylonian confusion" of the terms to describe it, ranging from mere constancy of crucial parameters (e.g. species composition or ecosystem processes) through a measure for the time a system resists to disturbances to resilience (see Brand and Jax 2007 for a discussion of the resilience concept). Also, the "ecosystem" is not something given by nature "as such", which has only to be identified, but also a matter of what is considered as an ecosystem. What makes up the "self-identity" of a system (Jax et al. 1998), so that we can speak of the "same" ecosystem even though things change within it (e.g. by birth and death, migration of organisms, or natural disturbances)? And when has it changed into another system, either a degraded one or an ecosystem of another type – or even ceases to be an ecosystem at all? These questions are far from trivial and academic. A (properly) functioning ecosystem may be characterised by the persistence of specific (native) species (often expressed as "ecosystem integrity"), or by the continued (sustainable) provision of ecosystem services, or by the maintenance of a particular physiognomy of an ecosystem (e.g. being a forest and not a savanna, even though species may have been replaced within the system, e.g. in novel ecosystems). A problem is that definitions are often not explicit. Thus, a physiognomically or simply spatially delimited area, such as a forest or a large wilderness area, is not only assumed to be an ecosystem of some (vaguely defined) kind. At the same time, particular assumptions on properties of the system and its functioning are made, without prior investigations. Such properties are commonly that the system is self-sustaining or self-regulating. This is often the case when, e.g., wilderness areas – as free from human intervention – are considered to be self-sustaining ("nature knows best"). Depending on the parameter one looks at, however, this may or may be not the case. Such untested assumptions can thus be problematic for the conservation or restoration of ecosystems (see Jax 2007). As will be explained in Section 4.2, this does not mean that "anything goes", that ecosystems are mere constructs of the human mind and that the question of their functioning is completely a matter of preferences and choices. But to have a clear idea of the targets of conservation and to operationalise concepts, such as ecosystem health and ecosystem integrity, not only empirical data and sound ecological theory are needed but also clear definitions and societal choices of what is in the focus of the conceptualisation of properly functioning ecosystems.

Conservation concepts compared: contrasts, links, and a matrix

My effort to sort the different conservation concepts by means of narratives, objects, and values turned out to be quite sobering. There are clear trends which differentiate the different conceptual clusters presented above, especially in terms of the narratives told about them. The narratives described in Section 3.3 are much less ambiguous than specific concepts discussed in Section 3.4, because the narratives are, partly exaggerated, ideal types of human–nature relationships. Differences and similarities are less clear with respect to the objects and values to which the major concepts (those naming each cluster) refer, especially when strategic considerations behind the use of the concepts are included into the picture. One then has to dive deeper into the different more specific meanings and interpretations of the terms denoting the major concepts, and into more specific expressions given by other terms of the respective cluster. The tables presented below are doing that for a set of concepts; taken together, these tables might serve as kind of matrix to visualise the characteristics of the concepts. Following the tables, some general contrasts and similarities between conservation concepts will be discussed.

The concepts listed in the tables are by necessity still a selection. I have tried to refer on the one hand to the most conspicuous and widely used concepts within each cluster and, secondly, to show the breadth of characteristics given by similar or seemingly identical concepts, as provided in the published literature and discussed earlier. There might be disagreement about the contents of the tables, but these tables are provided here only as a brief overview of my analysis of conservation concepts and not as the final "verdict" about the concepts. The elaboration of the history of the different concepts shows how difficult it is to really "fix" the meaning of particular concepts, given their ongoing dynamic developments. The tables are open to further discussion and refinement.

The first insight from this analysis is that most of the major conservation concepts, and even some more specific ones, are, at closer look, ambiguous with respect to the objects they address, the values they embrace, and – to a lesser degree – even to the narratives of human–nature relationships they are attached to. This insight may at first not appear to be very useful, but in fact it is. It is useful because quite often the concepts are discussed and communicated in a much too simplified way, not the least by those who oppose particular concepts and who set up strawmen of the concepts against which they then argue (see, e.g. the debate about the "new conservation" in Section 2.8). Being aware of this ambiguity is the first step to better reflect one's own aims and more clearly express one's own understanding of conservation and conservation concepts.

Some of the conceptual clusters are clearly programmatic in an action-oriented sense, especially restoration and ecosystem management, also cultural landscapes, as it refers to human interaction with non-human nature as a decisive characteristic. Other concepts at first sight seem to denote just objects or targets of conservation, namely wilderness, biodiversity, ecosystem services. These concepts, however, also stand for programmatic approaches and aim towards action.

There are many efforts in the literature to demonstrate that some of these major concepts are linked and reinforce each other, which in principle would be highly desirable for the conservation case: if you pursue one approach, you would also reach the goals of the other(s). Synergies to other major concepts are obvious for restoration, which, as described, builds on the other concepts. Conservationists also strive to demonstrate synergies between biodiversity and ecosystem services, as already mentioned earlier. There are certainly overlaps but given the complexities of both conceptual clusters, protecting ecosystem services does not necessarily mean protecting all biodiversity – and vice versa (Ridder 2008, Jax and Heink 2015). Likewise, close synergies have often been postulated between wilderness protection and biodiversity protection (e.g. recently Genes et al. 2019), a link that at least in that simple form has been rejected on good arguments (e.g. Sarkar 1999). Biodiversity protection does not necessarily exclude humans, but to the contrary, high biodiversity is often found in cultural landscapes shaped by long-lasting interactions between humans and non-human nature; sometimes biodiversity is even higher in such landscapes than in wild ones (see examples in Bridgewater and Rotherham 2019). It should appear as of no wonder that the clusters "cultural landscape" and "ecosystem services" are linked, but not to the degree that one simply can be considered as being able to replace the other. Historically, it is interesting to observe that concepts which at first aimed at the protection of biodiversity, such as rewilding and ecosystem management, over time then were also justified with and used for the protection of ecosystem services. These may have strategic reasons (e.g. making them "fit for policy", Pettorelli et al. 2018) or may be an extension of the respective concept for new (fashionable?) purposes – however useful in the end.

Overall legend for Tables 3.2–3.8

Narratives are abbreviated: **Apart**: Humans apart from nature, **Caring**: Caring for other living beings (and natural features) is part of a good human life, **Useful**: Safeguarding necessary and useful nature, **Home**: Being at home in nature, **Steering**: Steering nature in a responsible way, **Gardening**: Gardening nature in novel ways, **Whole again**: Making nature whole again. The *dominant* narrative (if there is) is indicated in bold.

Objects: Some terms, in particular ecosystem services and biodiversity, not only refer to programmatic approaches but also are used to denote specific objects (e.g.

Analysing conservation concepts 133

TABLE 3.2 Concepts within the cluster "ecosystem services" and their characteristics. Regarding the ultimate objects, "biodiversity" is often an implicit ultimate target, for which the concepts are used strategically

Cluster "Ecosystem services"

Concept	**Dominant narrative** and other narratives of importance	Objects a Pragmatic b Ultimate	Values embraced	Strategic dimensions
Ecosystem services	**Useful**, Steering, Home, Gardening, Caring	a Selected components (objects and processes) of ecosystems b Human well-being [biodiversity]	Instrumental, relational	High
Cultural services	**Useful**, Home, Caring		Relational, instrumental	High
Natural capital	**Useful**		Instrumental	High
Nature's contributions to people	**Useful**, Steering, Home, Gardening, Caring		Instrumental, relational	High
Nature-based solutions	Useful, Caring, Home		Instrumental, Relational	High

134 Analysing conservation concepts

TABLE 3.3 Concepts within the cluster "biodiversity" and their characteristics

Cluster "Biodiversity"

Concept	**Dominant narrative** and other narratives of importance	Objects a Pragmatic b Ultimate	Values embraced	Strategic dimensions
Biodiversity	**Caring**, Useful, Home, Steering, Apart	a Species, species richness, diversity, "typical" species, genetic diversity, ecosystems, nature as a whole b Biodiversity, human well-being	Intrinsic, relational, instrumental	High
Functional diversity	**Useful**, Steering	a Selected types of species b Ecosystems, human well-being	Instrumental	Medium
Natural monument	**Home**, Caring	a Selected natural features: individuals, abiotic features, species, landscapes b Human well-being	Relational	Low
Natural heritage	Caring, Home	a Selected natural features: individuals, abiotic features, species, landscapes b Human well-being	Relational	Medium

Analysing conservation concepts 135

TABLE 3.4 Concepts within the cluster "wilderness" and their characteristics

Cluster "Wilderness"

Concept	Dominant narrative and other narratives of importance	Objects a Pragmatic b Ultimate	Values embraced	Strategic dimensions
Wilderness	**Apart**, Caring, (Useful)	a *Wild* or *primeval* landscapes, ecosystems, ecosystem processes b (*Native*) species, populations, individuals, human experiences, human well-being	Intrinsic, relational, (instrumental)	Low
Wildness	Caring, Apart, Gardening	a Landscapes, individuals, populations, species, ecosystem processes b Species, populations, and individuals, human experiences, human well-being	Intrinsic, relational, (instrumental)	Low
Rewilding	Apart, Caring, Useful, Steering, Gardening	a Landscapes, individuals, populations, species, ecosystem processes, ecosystem services b Species, populations, individuals, human experiences, human well-being	Intrinsic, instrumental	Medium

136 Analysing conservation concepts

TABLE 3.5 Concepts within the cluster "cultural landscapes" and their characteristics

Cluster "Cultural landscapes"

Concept	**Dominant narrative** and other narratives of importance	Objects a Pragmatic b Ultimate	Values embraced	Strategic dimensions
Cultural landscapes	**Home**, Useful, Caring, Steering	a (traditional) Landscapes, species populations, ecosystem services, (management practices) b Human well-being, species	Relational, instrumental [intrinsic]	Medium
Semi-natural area	Caring, Useful, Steering, Home	a (traditional) Landscapes, species populations, human practices, ecosystem services, b Human well-being, species	Relational, instrumental, intrinsic	Low
Homeland/*Heimat*	**Home**, Caring	a Traditional landscapes b Human well-being	Relational	Medium
Natural heritage	Home, Caring	a Selected natural features: individuals, abiotic features, species, landscapes b Human well-being	Relational	Medium
Biocultural conservation	Home, Caring, Steering	a Traditional landscapes, species, ecosystem services b Human well-being, species	Relational, intrinsic, instrumental	Medium

Analysing conservation concepts **137**

TABLE 3.6 Concepts within the cluster "ecosystem management" and their characteristics

Cluster "Ecosystem management"

Concept	**Dominant narrative** and other narratives of importance	Objects a Pragmatic b Ultimate	Values embraced	Strategic dimensions
Ecosystem management	**Steering**, Caring, Useful	a Ecosystems, species populations, ecosystem services b Human well-being, species, biodiversity	Instrumental, intrinsic	Medium
Ecosystem-based management	**Steering**, Useful, Caring	a Ecosystems, ecosystems together with social systems, ecosystem services b Human well-being, biodiversity	Instrumental, intrinsic	Medium
Social-ecological system, socio-ecological systems	**Steering**, Useful, Caring, Home (?)	a Ecosystems including social systems b Human well-being, biodiversity	Instrumental, intrinsic, relational (?)	Medium
Ecological integrity	**Caring**, Steering, Useful	a Ecosystems, species b Biodiversity, human well-being	Intrinsic, instrumental	Medium
Ecosystem health	**Useful**, Caring, Steering	a Ecosystems, ecosystem services, species b Human well-being, biodiversity	Instrumental, intrinsic	High

138 Analysing conservation concepts

TABLE 3.7 Concepts within the cluster "novel ecosystems" and their characteristics. The concepts of "novel ecosystem" and those of "no-analogue" or "emerging" ecosystem are used largely synonymous, differ mainly in that the first may create more positive connotations than the others

Cluster "Novel ecosystems"

Concept	**Dominant narrative** and other narratives of importance	Objects a Pragmatic b Ultimate	Values embraced	Strategic dimensions
Novel ecosystem	**Gardening**, Caring, Useful	a Ecosystems, species populations b Biodiversity, human well-being	Intrinsic, instrumental, (relational)	Medium
No-analogue ecosystem/ emerging ecosystem	**Gardening**, Caring, Useful	a Ecosystems, species populations b Biodiversity, human well-being	Intrinsic, instrumental, (relational)	Low
Hybrid ecosystem	**Gardening**	Transient, depending on further development and goals	Transient, depending on further development and goals	Low
Designer ecosystem	**Useful**, Steering	a Ecosystems, landscapes, species, ecosystem goods and services b Human well-being	Instrumental, (relational)	Low

Analysing conservation concepts 139

TABLE 3.8 Concepts within the cluster "restoration" and their characteristics. Delimitations of the more specific concepts based on Gann et al. (2019)

Cluster "Restoration"

Concept	**Dominant narrative** and other narratives of importance	Objects a Pragmatic b Ultimate	Values embraced	Strategic dimensions
(ecological) Restoration	**Whole again**, Caring, Useful, Steering, Apart, Home	Depending on specific task and direction of restoration	Intrinsic, relational, instrumental	Medium
Reclamation	**Whole again**, Useful	a Ecosystems, ecosystem services b Human well-being	Instrumental	Low
Rehabilitation	**Whole again**, Useful, Steering	a Ecosystems, ecosystem services b Human well-being	Instrumental, relational	Low
Remediation	**Whole again**, (Caring, Useful, Steering, Apart, Home)	a Ecosystems, habitat conditions b Human health, *preparing for other types of restoration!*	Instrumental	Low

quantified ecosystem services, or biodiversity understood as species numbers). If "biodiversity" is named as an object, the term is meant to cover the full spectrum of meanings of biodiversity as an object. Special meanings are indicated, e.g., by "species richness", "genetic diversity", etc.

Values embraced: Only a rough classification is made here, i.e. between instrumental, relational, and intrinsic values. In those cases where several value types are mentioned, their order indicates also an order of importance.

Strategic dimension refers to the strategic use of a concept to address and shape conservation policy and public opinion.

Connections and overlaps between other conceptual clusters exist or have been postulated in various ways, but no strict convergences of objects, values, and narratives exist. There is thus no panacea, no unifying concept which may reach all conservation goals at the same time, addressing all conservation objects equally and serving the protection of all target objects and values embraced by conservationists. Choices between concepts remain necessary. Such choices should consider not only the contents of the respective conservation approach, i.e. what it includes, but also what is *excluded*. As conservation concepts are today discussed also in public, i.e., not only inside the conservation community but also outside of it, also the words matter, the connotations which words like "wilderness", "biodiversity", or "ecosystem services" carry with them (see also Chapter 5).

A narrative approach to conservation concepts, addressing the human–nature relationships involved, can be helpful for finding a way towards choosing appropriate concepts in specific settings. Being less ambiguous than the major conservation concepts, it may be easier to start with conservation narratives than to discuss time and again the "correct" meaning of major concepts. In addition, narratives come closer to the everyday life of people. In many controversies it may be the narratives that can foreground more sharply those issues in the use of conservation concepts that are in fact contested.

Some ideas for a way forward to selecting appropriate conservation concepts will be presented in Chapter 5, by necessity also including the social and economic context of conservation. Prior to that, however, it is necessary to first venture into some philosophical and anthropological background that is fundamental for all conservation narratives, conservation concepts, and conservation practice. This is the question of what "nature" and "natural" mean, and how different meanings of these concepts affect ideas of nature conservation. This is the subject of Chapter 4.

Notes

1 "die bewußten oder unbewußten Orientierungsstandards u. Leitvorstellungen, von denen sich Individuen u. Gruppen bei ihrer Handlungswahl leiten lassen".
2 The Ijsselmeer itself came into existence after the shallow Zuiderzee bay was closed towards the North Sea trough a 32-km-long dike in 1932.
3 Daily's space station as well as the "rivet popper" of the Ehrlichs' are metaphors. Using a metaphor means "understanding and experiencing one kind of thing in terms of another" (Lakoff and Johnson 2003, p. 5). Conservation concepts as well as conservation

narratives are replete with metaphors (see Larson 2011). A narrative can contain several metaphors and connect them.
4 I was inspired here by the classical works of Propp (2015/1958) on the "Morphology of the folk tale" and Levi-Strauss (1955) on the structure of myths.
5 All age data are meant as rough orientation, as orders of magnitude. I have not checked specific recent research, which might give some deviations from the age data mentioned.
6 Such a technique was also applied by Max Weber (1904/1988, p. 190ff) for creating "ideal types".
7 See, e.g. www.nps.gov/shen/learn/historyculture/displaced.htm. Accessed April 13, 2023.
8 A heritage, normally, is not just something you receive but also something you pass down to somebody else, mostly to the next generation.
9 "Steward". Merriam-Webster.com Dictionary, Merriam-Webster. www.merriam-webster.com/dictionary/steward. Accessed April 14, 2023.
10 www.cbd.int/convention/text/. Accessed May 1, 2023.
11 For a more detailed discussion of the rewilding concept see cluster "wilderness".

References

Abbott, H.P. (2008). *The Cambridge introduction to narrative*. 2nd ed. Cambridge University Press, Cambridge.
Allison, S.K. (2007). You can't not choose: embracing the role of choice in ecological restoration. *Restoration Ecology*, 15, 601–605.
Angermeier, P.L. (1994). Does biodiversity include artificial diversity? *Conservation Biology*, 8, 600–602.
Arias-Arévalo, P., Gómez-Baggethun, E., Martín-López, B. et al. (2018). Widening the evaluative space for ecosystem services: a taxonomy of plural values and valuation methods. *Environmental Values*, 27, 29–53.
Aronson, J., Dhillion, S. and Le Floc´h, E. (1995). On the need to select an ecosystem of reference, however imperfect: a reply to Pickett and Parker. *Restoration Ecology*, 3, 1–3.
Aronson, J., Floret, C., Le Floc´h, E. et al. (1993). Restoration and rehabilitation of degraded ecosystems in arid and semi-arid lands. I. A view from the south. *Restoration Ecology*, 1, 8–17.
Aronson, J., Milton, S.J. and Blignaut, J.N. (2007). Restoring natural capital: definitions and rationale. In: *Restoring natural capital: science, business, and practice* (eds. Aronson, J., Milton, S.J. and Blignaut, J.N.). Island Press, Washington D.C., pp. 3–8.
Ashby, W.C. (1987). Forests. In: *Restoration ecology* (eds. Jordan, W.R.I., Gilpin, M.E. and Aber, J.D.). Cambridge University Press, New York, pp. 89–108.
Bakken, P.W. (2009). Stewardship. In: *Encyclopedia of environmental ethics and philosophy* (eds. Callicott, J.B. and Frodeman, R.). MacMillan, Farmington Hills, Vol. 2, pp., 282–284.
Bal, M. (2017). *Narratology. Introduction to the theory of narrative*. University of Toronto Press, Toronto.
Bamberg, M. (2014). Identity and narration. In: *Handbook of narratology* (eds. Hühn, P., Meister, J.C., Pier, J. and Schmid, W.). Walter de Gruyter, Berlin, pp. 241–252.
Bennett, N.J., Whitty, T.S., Finkbeiner, E. et al. (2018). Environmental stewardship: a conceptual review and analytical framework. *Environmental Management*, 61, 597–614.
Berkes, F. and Folke, C. (eds.) (1998). *Linking social and ecological systems. Management practices and social mechanisms for building resilience*. Cambridge University Press, Cambridge.

Berr, K. and Schenk, W. (2019). Begriffsgeschichte. In: *Handbuch Landschaft* (eds. Kühne, O., Weber, F., Berr, K. and Jenal, C.). Springer, Wiesbaden, pp. 23–38.

Blignaut, J. and Aronson, J. (2020). Developing a restoration narrative: a pathway towards system-wide healing and a restorative culture. *Ecological Economics*, 168, 106483.

Bonneuil, C. (2015). The geological turn: narratives of the Anthropocene. In: *The Anthropocene and the global environmental crisis: rethinking modernity in a new epoch* (eds. Hamilton, C., Gemenne, F. and Bonneuil, C.). Routledge, London, pp. 17–31.

Boyce, M.S. and Haney, A. (eds.) (1997). *Ecosystem management. Applications for sustainable forest and wildlife resources*. Yale University Press, New Haven, London.

Brand, F.S. and Jax, K. (2007). Focussing the meaning(s) of resilience: resilience as a descriptive concept and a boundary object. *Ecology and Society*, 12(1): 23. www.ecologyandsociety.org/vol12/iss1/art23/

Bridgewater, P. and Rotherham, I.D. (2019). A critical perspective on the concept of biocultural diversity and its emerging role in nature and heritage conservation. *People and Nature*, 1, 291–304.

Brondizio, E.S., O'Brien, K., Bai, X. et al. (2016). Re-conceptualizing the Anthropocene: a call for collaboration. *Global Environmental Change*, 39, 318–327.

Brown, S. and Lugo, A.E. (1994). Rehabilitation of tropical lands: a key to sustaining development. *Restoration Ecology*, 2, 97–111.

Büscher, B. and Fletcher, R. (2020). *The conservation revolution. Radical ideas for saving nature beyond the Anthropocene*. Verso, London, New York.

Büscher, B., Fletcher, R., Brockington, D. et al. (2017). Half-Earth or whole Earth? Radical ideas for conservation, and their implications. *Oryx*, 51, 407–410.

Cadotte, M.W., Carscadden, K. and Mirotchnick, N. (2011). Beyond species: functional diversity and the maintenance of ecological processes and services. *Journal of Applied Ecology*, 48, 1079–1087.

Cafaro, P. and Primack, R. (2014). Species extinction is a great moral wrong. *Biological Conservation*, 170, 1–2.

Callicott, J.B. (2000a). Harmony between men and land – Aldo Leopold and the foundations of ecosystem management. *Journal of Forestry*, 98, 4–13.

Callicott, J.B. (2000b). Contemporary criticisms of the received wilderness idea. In: *Wilderness science in a time of change. Conference-Volume 1: Changing perspectives and future directions* (eds. Cole, D.N., McCool, S.F., Freimund, W.A. and O''Loughlin, J.). U.S. Department of Agriculture, Forest Service, Missoula, pp. 24–31.

Carson, R. (1962). *Silent spring*. Houghton Mifflin, Boston.

Chakrabarty, D. (2009). The climate of history: four theses. *Critical Inquiry*, 35, 197–222.

Chan, K.M.A., Balvanera, P., Benessaiah, K. et al. (2016). Why protect nature? Rethinking values and the environment. *Proceedings of the National Academy of Sciences of the United States of America*, 113, 1462–1465.

Chapin, F.S., Carpenter, S.R., Kofinas, G.P. et al. (2009). Ecosystem stewardship: sustainability strategies for a rapidly changing planet. *Trends in Ecology & Evolution*, 25, 241–249.

Clewell, A.F. and Aronson, J. (2006). Motivations for the restoration of ecosystems. *Conservation Biology*, 20, 420–428.

Coates, P. (2015). Creatures enshrined: wild animals as bearers of heritage. *Past & Present*, 226, 272–298.

Colding, J. and Barthel, S. (2019). Exploring the social-ecological systems discourse 20 years later. *Ecology and Society*, www.ecologyandsociety.org/vol24/iss1/art2/

Cole, D.N. (2000). Paradox of the primeval: ecological restoration in wilderness. *Ecological Restoration*, 18, 77–86.

Corlett, R.T. (2015). The Anthropocene concept in ecology and conservation. *Trends in Ecology and Evolution*, 30, 36–41.
Costanza, R. and Daly, H.E. (1992). Natural capital and sustainable development. *Conservation Biology*, 6, 37–46.
Costanza, R., d´Arge, R., de Groot, R. et al. (1997). The value of the world's ecosystem services and natural capital. *Nature*, 387, 253–260.
Craighead, F.C.J. (1979). *Track of the Grizzly*. Sierra Club, San Francisco.
Cronon, W. (1992). A place for stories: nature, history, and narrative. *Journal of American History*, 78, 1347–1376.
Crutzen, P.J. (2002). Geology of mankind. *Nature*, 415, 23.
Crutzen, P.J. and Stoermer, E.F. (2000). *Global Change Newsletter*, 41, 17–18.
Daily, G.C. (ed.) (1997). *Nature's services. Societal dependence on natural ecosystems*. Island Press, Washington D.C.
De Cesare, S. (2019). Disentangling organic and technological progress: an epistemological clarification introducing a key distinction between two levels of axiology. *Studies in History and Philosophy of Science Part C: Studies in History and Philosophy of Biological and Biomedical Sciences*, 73, 44–53.
Defries, R. and Nagendra, H. (2017). Ecosystem management as a wicked problem. *Science*, 356, 265–270.
Deliège, G. and Neuteleers, S. (2015). Should biodiversity be useful? Scope and limits of ecosystem services as an argument for biodiversity conservation. *Environmental Values*, 24, 165–182.
Díaz, S., Pascual, U., Stenseke, M. et al. (2018). Assessing nature's contributions to people. *Science*, 359, 270–272.
Dowie, M. (2009). *Conservation refugees. The hundred-year conflict between global conservation and native people*. MIT Press, Cambridge.
Ehrlich, P. and Ehrlich, A. (1981). *Extinction. The causes and consequences of the disappearance of species*. Random House, New York.
Elliot, R. (1982). Faking nature. *Inquiry (United Kingdom)*, 25, 81–93.
Elliott, J. (2005). *Using narrative in social research*. SAGE Publications, London.
Eser, U., Neureuther, A.-K., Seyfang, H. et al. (2014). *Prudence, justice and the good Life. A typology of reasoning in selected European national biodiversity strategies*. Bundesamt für Naturschutz, Bonn. https://portals.iucn.org/library/node/44639
Evers, C.R., Wardropper, C.B., Branoff, B. et al. (2018). The ecosystem services and biodiversity of novel ecosystems: a literature review. *Global Ecology and Conservation*, 13, e00362.
Flint, C.G., Kunze, I., Muhar, A. et al. (2013). Exploring empirical typologies of human-nature relationships and linkages to the ecosystem services concept. *Landscape and Urban Planning*, 120, 208–217.
Forman, R.T.T. and Godron, M. (1986). *Landscape ecology*. John Wiley & Sons, New York.
Fox, C.A., Reo, N.J., Turner, D.A. et al. (2017). "The river is us; the river is in our veins": re-defining river restoration in three Indigenous communities. *Sustainability Science*, 12, 521–533.
Fox, D. (2007). Back to the no-analog future? *Science*, 316, 823–824.
France, R.L. (ed.) (2008). *Healing natures, repairing relationships. New perspectives on restoring ecological spaces and consciousness*. Green Frigate Books, Sheffield.
Franzosi, R. (1998). Narrative analysis—or why (and how) sociologists should be interested in narrative. *Annual Review of Sociology*, 24, 517–554.

Fulda, D. (2014). Historiographic narration. In: *Handbook of narratology* (eds. Hühn, P., Meister, J.C., Pier, J. and Schmid, W.). Walter de Gruyter, Berlin, pp. 227–240.

Gann, G.D., McDonald, T., Walder, B. et al. (2019). International principles and standards for the practice of ecological restoration. Second edition. *Restoration Ecology*, 27, S1–S46.

Gardiner, S.M. and Thompson, A. (eds.) (2017). *The Oxford Handbook of Environmental Ethics*. Oxford University Press, Oxford.

Gaston, K.J. and Spicer, J.I. (2004). *Biodiversity: an introduction*. Blackwell, Oxford.

Gavin, M.C., McCarter, J., Mead, A. et al. (2015). Defining biocultural approaches to conservation. *Trends in Ecology & Evolution*, 30, 140–145.

Genes, L., Svenning, J.-C., Pires, A.S. et al. (2019). Why we should let rewilding be wild and biodiverse. *Biodiversity and Conservation*, 28, 1285–1289.

Gissibl, B., Höhler, S. and Kupper, P. (2012). Introduction. Towards a global history of national parks. In: *Civilizing nature. National Parks in global historical perspective* (eds. Gissibl, B., Höhler, S. and Kupper, P.). Berghahn Books, New York, pp. 1–27.

Gómez-Baggethun, E. and Ruiz-Pérez, M. (2011). Economic valuation and the commodification of ecosystem services. *Progress in Physical Geography*, 35, 613–628.

Gordon, I.J., Manning, A.D., Navarro, L.M. et al. (2021). Domestic livestock and rewilding: are they mutually exclusive? *Frontiers in Sustainable Food Systems*, 5. doi: 10.3389/fsufs.2021.550410

Görg, Christoph. (2016). Zwischen Tagesgeschäft und Erdgeschichte: Die unterschiedlichen Zeitskalen in der Debatte um das Anthropozän. *GAIA*, 25, 9–13.

Gorke, M. (2000). Was spricht für eine holistische Umweltethik? *Natur und Kultur*, 1, 86–105.

Gould, S.J. (1989). *Wonderful life. The Burgess shale and the nature of history*. Penguin, London.

Graf, R. and Jarusch, K.H. (2017). "Crisis" in contemporary history and historiography. In: *Docupedia Zeitgeschichte*. http://docupedia.de/zg/graf_jarausch_crisis_v1_en_2017; doi: 10.3389/fsufs.2021.550410

Grimm, V. and Wissel, C. (1997). Babel, or the ecological stability discussions: an inventory and analysis of terminology and a guide for avoiding confusion. *Oecologia*, 109, 323–334.

Grumbine, R.E. (1994). What is ecosystem management? *Conservation Biology*, 8, 27–38.

Guha, R. (1989). Radical American environmentalism and wilderness preservation: a third world critique. *Environmental Ethics*, 11, 71–83.

Haila, Y. (1997). 'Wilderness' and the multiple layers of environmental thought. *Environment and History*, 3, 129–147.

Hall, M. (2005). *Earth repair. A transatlantic history of environmental restoration*. University of Virginia Press, Charlottesville.

Head, L. (2014). Contingencies of the Anthropocene: lessons from the 'neolithic'. *The Anthropocene Review*, 1, 113–125.

Hertog, I.M. and Turnhout, E. (2018). Ideals and pragmatism in the justification of ecological restoration. *Restoration Ecology*, 26, 1221–1229.

Hettinger, N. (2005). Respecting nature's autonomy in relationship with humanity. In: *Recognizing the autonomy of nature: theory and practice* (ed. Heyd, T.). Columbia University Press, New York, pp. 86–98.

Higgs, E. (2017). Novel and designed ecosystems. *Restoration Ecology*, 25, 8–13.

Higgs, E., Falk, D.A., Guerrini, A. et al. (2014). The changing role of history in restoration ecology. *Frontiers in Ecology and the Environment*, 12, 499–506.

Higgs, E.S. (2003). *Nature by design. People, natural process, and ecological restoration*. MIT Press, Cambridge.

Higgs, E.S. (2013). Perspective: a tale of two natures. In: *Novel ecosystems. Intervening in the new ecological world order* (eds. Hobbs, R.J., Higgs, E.S. and Hall, C.M.). Wiley & Sons, New York, pp. 293–295.

Himes, A. and Muraca, B. (2018). Relational values: the key to pluralistic valuation of ecosystem services. *Current Opinion in Environmental Sustainability*, 35, 1–7.

Hoag, R.W. (1982). The mark on the wilderness: Thoreau's contact with Ktaadn. *Texas Studies in Literature and Language*, 24, 23–46.

Hobbs, R.J. (2016). Degraded or just different? Perceptions and value judgements in restoration decisions. *Restoration Ecology*, 24, 153–158.

Hobbs, R.J., Cole, D.N., Yung, L. et al. (2010). Guiding concepts for park and wilderness stewardship in an era of global environmental change. *Frontiers in Ecology and the Environment*, 8, 483–490.

Hobbs, R.J. and Harris, J.A. (2001). Restoration ecology: repairing the Earth's ecosystems in the new millennium. *Restoration Ecology*, 9, 239–246.

Hobbs, R.J., Higgs, E., Hall, C.M. et al. (2014). Managing the whole landscape: historical, hybrid, and novel ecosystems. *Frontiers in Ecology and the Environment*, 12, 557–564.

Hobbs, R.J., Higgs, E. and Harris, J.A. (2009). Novel ecosystems: implications for conservation and restoration. *Trends in Ecology & Evolution*, 24, 599–605.

Hobbs, R.J., Higgs, E.S. and Hall, C.M. (2013a). *Novel ecosystems: intervening in the new ecological world order*. Wiley & Sons, New York.

Hobbs, R.J., Higgs, E.S. and Hall, C.M. (2013b). Introduction: why novel ecosystems? In: *Novel ecosystems: Intervening in the new ecological world order* (eds. Hobbs, R.J., Higgs, E. and Hall, C.M.). Wiley & Sons, New York, pp. 1–8.

Horn, C. (1997). Wert. In: *Lexikon der Ethik* (ed. Höffe, O.). Beck, München, pp. 332–334.

Hühn, P., Meister, J.C., Pier, J. et al. (eds.) (2014). *Handbook of narratology*. Walter de Gruyter, Berlin.

Hull, D.L. (1997). The ideal species concept – and why we can't get it. In: *Species: the units of biodiversity* (eds. Claridge, M.F., Dawah, H.A. and Wilson, M.R.). Chapman & Hall, London, pp. 357–380.

Ingram, M., Ingram, H. and Lejano, R. (2015). Environmental action in the Anthropocene: the power of narrative networks. *Journal of Environmental Policy & Planning*, 21, 492–503.

Inkpen, S.A. (2017). Demarcating nature, defining ecology: creating a rationale for the study of nature's "primitive conditions". *Perspectives on Science*, 25, 355–392.

Jacques, D. (1995). The rise of cultural landscapes. *International Journal of Heritage Studies*, 1, 91–101.

Jannidis, F. (2014). Character. In: *Handbook of narratology* (eds. Hühn, P., Meister, J.C., Pier, J. and Schmid, W.). Walter de Gruyter, Berlin, pp. 30–45.

Jax, K. (2006). The units of ecology. Definitions and application. *Quarterly Review of Biology*, 81, 237–258.

Jax, K. (2007). Can we define ecosystems? On the confusion between definition and description of ecological concepts. *Acta Biotheoretica*, 55, 341–355.

Jax, K. (2010). *Ecosystem functioning*. Cambridge University Press, Cambridge.

Jax, K. and Heink, U. (2015). Searching for the place of biodiversity in the ecosystem services discourse. *Biological Conservation*, 191, 198–205.

Jax, K., Jones, C.G. and Pickett, S.T.A. (1998). The self-identity of ecological units. *Oikos*, 82, 253–264.

Jax, K. and Schwarz, A. (2011). Structure of the handbook. In: *Ecology revisited: reflecting on concepts, advancing science* (eds. Schwarz, A.E. and Jax, K.). Springer, Dordrecht, pp. 11–17.

Jones, M. (2003). The concept of cultural landscape: discourse and narrative. In: *Landscape interfaces* (eds. Palang, H. and Fry, G.). Kluwer Dordrecht, pp. 21–51.

Jordan, W.R., Gilpin, M.E. and Aber, J.D. (1987). Restoration ecology: ecological restoration as a technique to basic research. In: *Restoration ecology: a synthetic approach to ecological research* (eds. Jordan, W.R., Gilpin, M.E. and Aber, J.D.). Cambridge University Press, Cambridge, pp. 3–21.

Jordan, W.R.I. and Lubick, G.M. (2011). *Making nature whole: a history of ecological restoration*. Island Press, Washington D.C.

Jørgensen, D. (2015). Rethinking rewilding. *Geoforum*, 65, 482–488.

Journet, D. (1991). Ecological theories as cultural narratives – F.E. Clements's and H.A. Gleason's "stories" of community succession. *Written Communication*, 8, 446–472.

Kay, C.E. (1995). Aboriginal overkill and native burning. Implications for modern ecosystem management. In: *Sustainable society and protected areas: contributed papers of the 8th Conference on research and resource management in parks and on public lands, April 17–21, 1995, Portland, Oregon* (ed. Linn, R.). The George Wright Society, Hancock, MI, pp. 107–118.

Keeling, P. (2013). Wilderness, people, and the false charge of misanthropy. *Environmental Ethics*, 35, 387–405.

Kellert, S.R. (1996). *The value of life: biological diversity and human society*. Island Press, Washington D.C.

Kenter, J.O. (2018). IPBES: don't throw out the baby whilst keeping the bathwater; put people's values central, not nature's contributions. *Ecosystem Services*, 33, 40–43.

Kenter, J.O., O'Brien, L., Hockley, N. et al. (2015). What are shared and social values of ecosystems? *Ecological Economics*, 111, 86–99.

King, E.G. and Hobbs, R.J. (2006). Identifying linkages among conceptual models of ecosystem degradation and restoration: towards an integrative framework. *Restoration Ecology*, 14, 369–378.

Kirchhoff, T. (2019). Abandoning the concept of cultural ecosystem services, or against natural – scientific imperialism. *Bioscience*, 69, 220–227.

Kirchhoff, T., Brand, F. and Hoheisel, D. (2012). From cultural landscapes to resilient social-ecological systems: transformation of a classical paradigm or novel approach? In: *Resilience and the cultural landscape* (eds. Plieninger, T. and Bieling, C.). Cambridge University Press, Cambridge, pp. 49–64.

Kirchhoff, T., Trepl, L. and Vicenzotti, V. (2013). What is landscape ecology? An analysis and evaluation of six different conceptions. *Landscape Research*, 38, 33–51.

Kirchhoff, T. and Vicenzotti, V. (2014). A historical and systematic survey of European perceptions of wilderness. *Environmental Values*, 23, 443–464.

Klaver, I., Keulartz, J., van den Belt, H. et al. (2002). Born to be wild: a pluralistic ethics concerning introduced large herbivores in the Netherlands. *Environmental Ethics*, 24, 3–21.

Kohen, J.L. (2003). Knowing country: indigenous Australians and the land. In: *Nature across cultures: views of nature and the environment in non-Western cultures* (ed. Selin, H.). Springer, Dordrecht, pp. 229–243.

Kopnina, H., Leadbeater, S.R.B. and Cryer, P. (2019). Learning to rewild: examining the failed case of the Dutch "New Wilderness" Oostvaardersplassen. *International Journal of Wilderness*, 25, 72–89.

Koselleck, R. (2002). Some questions regarding the conceptual history of "crisis". In: *The practice of conceptual history: timing history, spacing concepts* (ed. Koselleck, R.). Stanford University Press, Stanford, pp. 236–247.

Koselleck, R. and Richter, M.W. (2006). Crisis. *Journal of the History of Ideas*, 67, 357–400.

Lackey, R.T. (1998). Seven pillars of ecosystem management. *Landscape and Urban Planning*, 40, 21–30.

Lakoff, G. and Johnson, M. (2003). *Metaphors we live by*. University of Chicago Press, Chicago, London.

Lange, M.M. (2019). Progress. In: *The Stanford encyclopedia of philosophy (Winter 2019 Edition)* (ed. Zalta, E.N.). https://plato.stanford.edu/entries/progress/

Larson, B.M.H. (2011). *Metaphors for environmental sustainability: redefining our relationship with nature*. Yale University Press, New Haven.

Lejano, R., Ingram, M. and Ingram, H. (2013). *The power of narrative in environmental networks*. MIT Press, Cambridge.

Leopold, A. (1949). *A Sand County almanac and sketches here and there*. Oxford University Press, New York.

Leopold, A. (1993). The round river: a parable. In: *Round river: from the journals of Aldo Leopold* (ed. Leopold, L.). Oxford University Press, Oxford, pp. 158–165.

Leopold, A.S., Cain, S.A., Cottham, C.M. et al. (1963). Wildlife management in the national parks. *Transactions of the North American Wildlife and Natural Resources Conference*, 28, 28–45.

Lévi-Strauss, C. (1955). The structural study of myth. *The Journal of American Folklore*, 68, 428–444.

Lewis, M. (ed.) (2007). *American wilderness: a new history*. Oxford University Press, Oxford, New York.

Lewis, S.L. and Maslin, M.A. (2015). Defining the Anthropocene. *Nature*, 519, 171–180.

Lindenmayer, D.B., Fischer, J., Felton, A. et al. (2008). Novel ecosystems resulting from landscape transformation create dilemmas for modern conservation practice. *Conservation Letters*, 1, 129–135.

Liszka, J.J. (2003). The narrative ethics of Leopold's Sand County Almanac. *Ethics and the Environment*, 8, 42–70.

Lorimer, J. (2015). *Wildlife in the Anthropocene: conservation after nature*. University of Minnesota Press, Minneapolis, London.

Lorimer, J. and Driessen, C. (2014). Wild experiments at the Oostvaardersplassen: rethinking environmentalism in the Anthropocene. *Transactions of the Institute of British Geographers*, 39, 169–181.

Lorimer, J., Sandom, C., Jepson, P. et al. (2015). Rewilding: science, practice, and politics. *Annual Review of Environment and Resources*, 40, 39–62.

Louder, E. and Wyborn, C. (2020). Biodiversity narratives: stories of the evolving conservation landscape. *Environmental Conservation*, 47, 251–259.

Mace, G., Norris, K. and Fitter, A.H. (2012). Biodiversity and ecosystem services: a multilayered relationship. *Trends in Ecology & Evolution*, 27, 19–26.

Maclaurin, J. and Sterelny, K. (2008). *What is biodiversity?* University of Chicago Press, Chicago.

Maier, D.S. (2013). *What's so good about biodiversity? A call for better reasoning about nature's value*. Springer, Dordrecht.

Malm, A. and Hornborg, A. (2014). The geology of mankind? A critique of the Anthropocene narrative. *The Anthropocene Review*, 1, 62–69.

Marris, E. (2011). *Rambunctious garden: saving nature in a post-wild world.* Bloomsbury, New York.
Marris, E. (2015). Humility in the Anthropocene. In: *After preservation: saving American nature in the age of humans* (eds. Minteer, B.A. and Pyne, S.J.). University of Chicago Press, Chicago, pp. 41–49.
Marris, E., Mascaro, J. and Ellis, E.C. (2013). Perspective: is everything a novel ecosystem? If so, do we need the concept? In: *Novel ecosystems: intervening in the new ecological world order* (eds. Hobbs, R.J., Higgs, E.S. and Hall, C.M.). Wiley & Sons, New York, pp. 345–349.
Martin, L.J. (2022). *Wild by design: the rise of ecological restoration.* Harvard University Press, Cambridge.
McCord, E.L. (2012). *The value of species.* Yale University Press, New Haven, London.
McEuen, A.B. and Styles, M.A. (2019). Is gardening a useful metaphor for conservation and restoration? History and controversy. *Restoration Ecology*, 27, 1194–1198.
Meadows, D.H., Meadows, D.L., Randers, J. et al. (1972). *Limits to growth: a report for the Club of Rome's project on the predicament of mankind.* Potomac Associates, London.
Medawar, P. (1982). *Pluto's republic.* Oxford University Press, Oxford, New York.
Meine, C. (2020). From the land to socio-ecological systems: the continuing influence of Aldo Leopold. *Socio-Ecological Practice Research*, 2, 31–38.
Meuter, N. (2014). Narration in various disciplines. In: *Handbook of narratology* (eds. Hühn, P., Meister, J.C ., Pier, J. and Schmid, W.). Walter de Gruyter, Berlin, pp. 447–467.
Meyer-Abich, K.M. (1984). *Wege zum Frieden mit der Natur. Praktische Naturphilosophie für die Umweltpolitik.* Carl Hanser, München.
Millennium Ecosystem Assessment (2005). *Ecosystems and human well-being: synthesis.* Island Press, Washington D.C.
Miller, J.R. and Bestelmeyer, B.T. (2016). What's wrong with novel ecosystems, really? *Restoration Ecology*, 24, 577–582.
Moreira, F., Queiroz, A.I. and Aronson, J. (2006). Restoration principles applied to cultural landscapes. *Journal for Nature Conservation*, 14, 217–224.
Morse, N.B., Pellissier, P.A., Cianciola, E.N. et al. (2014). Novel ecosystems in the Anthropocene: a revision of the novel ecosystem concept for pragmatic applications. *Ecology and Society*, 19(2), 12. http://dx.doi.org/10.5751/ES-06192-190212.
Muraca, B. (2011). The map of moral significance: a new axiological matrix for environmental ethics. *Environmental Values*, 20, 375–396.
Muradian, R. and Gómez-Baggethun, E. (2021). Beyond ecosystem services and nature's contributions: is it time to leave utilitarian environmentalism behind? *Ecological Economics*, 185, 107038.
Muradian, R. and Pascual, U. (2018). A typology of elementary forms of human-nature relations: a contribution to the valuation debate. *Current Opinion in Environmental Sustainability*, 35, 8–14.
Murcia, C., Aronson, J., Kattan, G.H. et al. (2014). A critique of the 'novel ecosystem' concept. *Trends in Ecology and Evolution*, 29, 548–553.
Narayan, D., Chambers, R., Shah, M.K. et al. (2000). *Voices of the poor: crying out for change.* Oxford University Press, New York.
Nash, R.F. (2014). *Wilderness and the American mind.* 5th ed. Yale University Press, New Haven.
Navarro, L.M. and Pereira, H.M. (2012). Rewilding abandoned landscapes in Europe. *Ecosystems*, 15, 900–912.

Nelson, M.P. (1998). An amalgamation of wilderness preservation arguments. In: *The great new wilderness debate* (eds. Callicott, J.B. and Nelson, M.P.). University of Georgia Press, Athens, pp. 154–198.

Nesshöver, C., Assmuth, T., Irvine, K.N. et al. (2017). The science, policy and practice of nature-based solutions: an interdisciplinary perspective. *Science of the Total Environment*, 579, 1215–1227.

Norton, B.G. (1987). *Why preserve natural variety?* Princeton University Press, Princeton, New Jersey.

Norton, B.G. (2002). Caring for nature: a broader look at animal stewardship. In: *Searching for sustainability: interdisciplinary essays in the philosophy of conservation biology* (ed. Norton, B.G.). Cambridge University Press, Cambridge, pp. 375–395.

O'Neill, J., Holland, A. and Light, A. (2008). *Environmental values*. Routledge, London.

Oelschlaeger, M. (1991). *The idea of wilderness*. Yale University Press, New Haven, London.

Olwig, K.R. (1996). Recovering the substantive nature of landscape. *Annals of the Association of American Geographers*, 86, 630–653.

Partelow, S. (2018). A review of the social-ecological systems framework: applications, methods, modifications, and challenges. *Ecology and Society*, 23. www.ecologyandsociety.org/vol23/iss4/art36/

Pascual, U., Balvanera, P., Christie, M. et al. (eds.) (2022). *Summary for policymakers of the methodological assessment report on the diverse values and valuation of nature of the Intergovernmental Science-Policy Platform on Biodiversity and Ecosystem Services.* IPBES secretariat, Bonn.

Pascual, U., Balvanera, P., Díaz, S. et al. (2017). Valuing nature's contributions to people: the IPBES approach. *Current Opinion in Environmental Sustainability*, 26–27, 7–16.

Perring, M.P., Standish, R.J., Price, J.N. et al. (2015). Advances in restoration ecology: rising to the challenges of the coming decades. *Ecosphere*, 6, art 131.

Peterson, R.O. (1995). *The wolves of Isle Royale: a broken balance*. Willow Creek Press, Minocqua, Wisconsin.

Peterson, R.O. (1999). Wolf-moose interaction on Isle Royale: the end of natural regulation? *Ecological Applications*, 9, 10–16.

Pettorelli, N., Barlow, J., Stephens, P.A. et al. (2018). Making rewilding fit for policy. *Journal of Applied Ecology*, 55, 1114–1125.

Phillips, A. (1998). The nature of cultural landscapes – a nature conservation perspective. *Landscape Research*, 23, 21–38.

Pickett, S.T.A., Parker, V.T. and Fiedler, P.L. (1992). The new paradigm in ecology: implications for conservation biology above the species level. In: *Conservation biology: the theory and practice of conservation, preservation and management* (eds. Fiedler, P.L. and Jain, S.K.). Chapman & Hall, New York, pp. 65–88.

Potthoff, K. (2013). The use of 'cultural landscape' in 19th century German geographical literature. *Norsk Geografisk Tidsskrift – Norwegian Journal of Geography*, 67, 49–54.

Propp, V. (2015/1958). *Morphology of the folktale*. Martino Publishing, Mansfield Centre, Connecticut.

Rapp, F. (1992). *Fortschritt. Entwicklung und Sinngehalt einer philosophischen Idee*. Wissenschaftliche Buchgesellschaft, Darmstadt.

Redford, K.H. and Adams, W.M. (2009). Payments for ecosystem services and the challenge of saving nature. *Conservation Biology*, 23, 785–787.

Reich, J. (2001). Re-creating the wilderness: shaping narratives and landscapes in Shenandoah National Park. *Environmental History*, 6, 95–117.

Ricoeur, P. (1992). Narrative identity. In: *On Paul Ricoeur: narrative and interpretation* (ed. Wood, D.). Routledge, London, pp. 188–199.

Ridder, B. (2008). Questioning the ecosystem services argument for biodiversity conservation. *Biodiversity and Conservation*, 17, 781–790.

Rissman, A.R. and Gillon, S. (2017). Where are ecology and biodiversity in social–ecological systems research? A review of research methods and applied recommendations. *Conservation Letters*, 10, 86–93.

Robertson, M., Nichols, P., Horwitz, P. et al. (2000). Environmental narratives and the need for multiple perspectives to restore degraded landscapes in Australia. *Ecosystem Health*, 6, 119–133.

Rozzi, R. (1999). The reciprocal links between evolutionary-ecological sciences and environmental ethics. *BioScience*, 49, 911–921.

Rozzi, R. (2013). Biocultural ethics: from biocultural homogenization toward biocultural conservation. In: *Linking ecology and ethics for a changing world: values, philosophy, and action* (eds. Rozzi, R., Pickett, S.T.A., Palmer, C., Armesto, J.J. and Callicott, J.B.). Springer, Dordrecht, pp. 9–32.

Rozzi, R. (2015a). Earth Stewardship and the biocultural ethic: Latin American perspectives. In: *Earth stewardship: linking ecology and ethics in theory and practice* (eds. Rozzi, R., Chapin, F.S.I., Callicott, J.B. et al.). Springer, Heidelberg, pp. 87–112.

Rozzi, R. (2015b). Implications of the biocultural ethic for Earth stewardship. In: *Earth stewardship: linking ecology and ethics in theory and practice* (eds. Rozzi, R., Chapin, F.S.I., Callicott, J.B. et al.). Springer, Heidelberg, pp. 113–136.

Rozzi, R., Chapin, F.S.I., Callicott, J.B. et al. (eds.) (2015). *Earth stewardship: linking ecology and ethics in theory and practice*. Springer, Heidelberg.

Ryan, J. (2012). Narrative environmental ethics, nature writing, and ecological science as tradition: towards a sponsoring ground of concern. *Philosophy Study*, 2, 822–843.

Ryan, M.-L. (2007). Toward a definition of narrative. In: *The Cambridge companion to narrative* (ed. Herman, D.). Cambridge University Press, Cambridge, pp. 22–35.

Sarkar, S. (1999). Wilderness preservation and biodiversity conservation – keeping divergent goals distinct. *BioScience*, 49, 405–412.

Schama, S. (1995). *Landscape and memory*. Knopf, New York.

Scherzinger, W. (1990). Das Dynamik-Konzept im flächenhaften Naturschutz – Zieldiskussion am Beispiel der Nationalpark-Idee. *Natur und Landschaft*, 65, 292–298.

Schlaepfer, M.A., Sax, D.F. and Olden, J.D. (2011). The potential conservation value of non-native species. *Conservation Biology*, 25, 428–437.

Schmelzer, M. (2015). The growth paradigm: history, hegemony, and the contested making of economic growthmanship. *Ecological Economics*, 118, 262–271.

Schwartz, D.T. (2019). European experiments in rewilding: Oostvaardersplassen. *The Rewilding Institute*, https://rewilding.org/european-experiments-in-rewilding-oostvaardersplassen/

Schwarz, A. (2011). History of concepts for ecology. In: *Ecology revisited: reflecting on concepts, advancing science* (eds. Schwarz, A. and Jax, K.). Springer, Dordrecht, pp. 19–28.

Scott, D.W. (2001/2002). "Untrammeled", "wilderness character" and the challenges of wilderness preservation. *Wild Earth*, 11, 72–79.

Seefried, E. (2015). Rethinking progress. On the origin of the modern sustainability discourse, 1970–2000. *Journal of Modern European History*, 13, 377–400.

Solbrig, O.T. (1991). *Biodiversity: scientific issues and collaborative research proposals*. UNESCO, Paris.

Soulé, M. (2013). The "new conservation". *Conservation Biology*, 27, 895–897.
Soulé, M. and Noss, R. (1998). Rewilding and biodiversity: complementary goals for continental conservation. *Wild Earth*, 8, 18–28.
Soulé, M.E. (1985). What is conservation biology? *BioScience*, 35, 727–734.
Soulé, M.E. (1996). Are ecosystem processes enough? *Wild Earth*, 6, 56–59.
Spence, M.D. (1999). *Dispossessing the wilderness. Indian removal and the making of the national parks*. Oxford University Press, New York, Oxford.
Standish, R.J., Thompson, A., Higgs, E.S. et al. (2013). Concerns about novel ecosystems. In: *Novel ecosystems: intervening in the new ecological world order* (eds. Hobbs, R.J., Higgs, E.S. and Hall, C.M.). Wiley & Sons, New York, pp. 296–308.
Stanley, T.R.J. (1995). Ecosystem management and the arrogance of humanism. *Conservation Biology*, 9, 255–262.
Sysling, F. (2015). 'Protecting the primitive natives': indigenous people as endangered species in the early nature protection movement, 1900–1940. *Environment and History*, 21, 381–399.
Takacs, D. (1996). *The idea of biodiversity: philosophies of paradise*. John Hopkins University Press, Baltimore, London.
Taylor, K. (2008). Landscape and memory: cultural landscapes, intangible values and some thoughts on Asia. In: *16th ICOMOS General Assembly and International Symposium: 'Finding the spirit of place – between the tangible and the intangible', 29 Sept–4 Oct 2008, Quebec, Canada*. Australia Memory of the World programme, online, pp. 1–6.
Taylor, K. and Lennon, J. (2011). Cultural landscapes: a bridge between culture and nature? *International Journal of Heritage Studies*, 17, 537–554.
Taylor, K. and Lennon, J.L. (2012). *Managing cultural landscapes*. Routledge, London.
Theunissen, B. (2019). The Ostvaardersplassen fiasco. *Isis*, 110, 341–345.
Thoreau, H.D. (1862/1992). *Walking*. Applewood Books, Bedford, Massachusetts.
Thoreau, H.D. (1864/1983). *The Maine woods*. Princeton University Press, Princeton, New Jersey.
Thornton, T.F. and Thornton, P.M. (2015). The mutable, the mythical, and the managerial: raven narratives and the Anthropocene. *Environment and Society: Advances in Research*, 6, 66–86.
Treanor, B. (2008). Narrative environmental virtue ethics: phronesis without a phronimos. *Environmental Ethics*, 30, 361–379.
Trepl, L. (2012). *Die Idee der Landschaft. Eine Kulturgeschichte von der Aufklärung bis zur Ökologiebewegung*. transcript Verlag, Bielefeld.
Truitt, A.M., Granek, E.F., Duveneck, M.J. et al. (2015). What is novel about novel ecosystems: managing change in an ever-changing world. *Environmental Management*, 55, 1217–1226.
UNEP/CBD (2000). Ecosystem approach. In: *Decisions adopted by the Conference of the Parties to the Convention on Biological Diversity at its Fifth Meeting, Nairobi, 15–26 May 2000*. www.cbd.int/decision/cop/?id=7148. Accessed March 10, 2021.
UNESCO (2021). *Operational guidelines for the implementation of the World Heritage Convention*. UNESCO, Paris.
van den Belt, H. (2004). Networking nature, or Serengeti behind the dikes. *History and Technology*, 20, 311–333.
van Dyke, F. and Lamb, R.L. (2020). *Conservation biology: foundations, concepts, applications*. Springer, Cham.

van Wieren, G. (2008). Ecological restoration as public spiritual practice. *Worldviews: Global Religions, Culture, and Ecology*, 12, 237–254.
Veldman, R.G. (2012). Narrating the environmental apocalypse: how imagining the end facilitates moral reasoning among environmental activists. *Ethics and the Environment*, 17, 1–23.
Vera, F.W.M. (2009). Large-scale nature development – the Oostvaardersplassen. *British Wildlife*, 20, 28–36.
von Droste, B., Plachter, H. and Rössler, M. (eds.) (1995). *Cultural landscapes of universal value*. Gustav Fischer, Jena, Stuttgart, New York.
Weber, M. (1904/1988). Die "Objektivität" sozialwissenschaftlicher und sozialpolitischer Erkenntnis. In: *Gesammelte Aufsätze zur Wissenschaftslehre*. J.C.B. Mohr Tübingen, pp. 146–214.
West, S., Jamila, H.L., Vanessa, M. et al. (2018). Stewardship, care and relational values. *Current Opinion in Environmental Sustainability*, 35, 30–38.
Westman, W.E. (1977). How much are nature's services worth? *Science*, 197, 960–964.
White, H. (1980). The value of narrativity in the representation of reality. *Critical Inquiry*, 7, 5–27.
Wilson, E.O. (ed.) (1988). *Biodiversity*. National Academy Press, Washington D.C.
Wilson, E.O. (2016). *Half Earth: our planet's fight for life*. Liveright Publication Corporation, New York.
Wuerthner, G., Crist, E. and Butler, T. (eds.) (2014). *Keeping the wild: against the domestication of the earth*. Island Press, Washington D.C.

4
WESTERN AND NON-WESTERN IDEAS OF NATURE AND NATURE CONSERVATION

This chapter deals with various ideas of nature that are contained in conservation concepts. Already the titles of several books demonstrate that "nature" and its specific meanings are far more than only of academic interest. William Cronon's famous book *Uncommon Ground*, e.g., asked in its subtitle for "reinventing nature" (first printing 1995) and in the paperback printing, one year later, at least for "rethinking the human place in nature". The book was met by strong criticism, as many conservationists considered it as an assault on the very idea of nature and the natural, and in particular on wilderness, as one important embodiment of the "natural" in conservation. In the same year Soulé and Lease (1995) published a book entitled "Reinventing nature? Responses to postmodern deconstruction", which might be seen as a reaction to approaches such as those of Cronon and his colleagues. Other books relating conservation to ideas of nature are, e.g., Cole and Young's *Beyond Naturalness: Rethinking Parks and Wilderness Stewardship in an Era of Rapid Change* (2010) or Lorimer's *Wildlife in the Anthropocene: Conservation after Nature* (2015). They all show the persistent importance of the idea of "nature" and "the natural" in conservation.

To approach this issue it is necessary to first delve into some philosophical and anthropological discourses on the idea of nature and the natural, and how they figure in conservation debates (Section 4.1). I will then describe and analyse the debate about whether nature (or parts of it) is considered to be a human construct or if it is real in a fundamental sense (Section 4.2). Section 4.3 will reach beyond the usual Western ideas of nature and look at how nature is (if at all) conceptualised in other cultures and traditions. Section 4.4 will illustrate the difficulties to "translate" between Western, scientific ideas and those from other cultures – using the example of the IPBES Conceptual Framework. While narrative analysis will be referred to,

DOI: 10.4324/9781003251002-4

occasionally, already in Sections 4.1–4.4, the usefulness of narratives of human–nature relationships for linking Western and non-Western approaches to nature will be dealt with in some more detail in Section 4.5.

4.1 What is nature? Philosophical and anthropological discourses

"Nature" characterises a relation between humans and the physical world. Within the Western world, different concepts and different understandings of nature exist since antiquity. The history of the concept of nature has been described by several authors (e.g. Williams 1980, Hager et al. 1984, Coates 1998). This history has a persisting influence on today's perception of nature and the natural. The different ideas shaped human relationships to the world (see Glacken 1967), but they also were and are crucial for the self-understanding of humans.

Different dichotomies characterising nature in Western philosophy

Most Western definitions of "nature" were formulated in a dualistic manner, as being in opposition to some other concept – what nature is *not*. It is by these contrasts that the term "nature" derives its specific meanings. Philosopher Dieter Birnbacher therefore speaks of the concept of the "natural" as a "semantic chameleon":

> They adapt their colour to the respective environment. Each time when the term "natural" is used, it is to foreground a contrast and to distinguish between the natural and its respective opposite.
> *(Birnbacher 2006, p. 6, translation KJ)*[1]

The most general and one of the oldest understandings is that of nature as the **totality of being**, as everything that exists in the biophysical world. At the first sight this might not be a dualistic idea, as it also includes humans, all their actions, and even their artifices. The opposite of "nature", however, is the "supernatural" in this case, the realm of the divine or the spiritual. This understanding of nature is basic to modern science and related to the notion of the "laws of nature". Describing the "natural" in this manner, however, is not the main idea of nature and naturalness that underlies most conservation approaches.

Another understanding of the word "nature" is that it denotes the **essence of something**. We commonly speak about "the nature of" something, meaning what characterises this object, i.e. "its very nature" (e.g. it is the nature of humans to speak and it is the nature of fish to swim). The opposing term in this case is the "unnatural". What counts as "unnatural" (e.g. a particular behaviour or state of something) is, however, often difficult to decide.

Beyond these definitions, most other definitions create a dichotomy between something human and **what is not human**. On behalf of this, the philosopher

Robert Spaemann – in a philosophical dictionary – characterises the origins of these dichotomies and the ensuing concepts not as being philosophical but originally deriving from "human practice" (Spaemann 1973, p. 957). The classical oppositions, still valid today, are nature–humans, nature–culture, nature–society, nature–artifice, nature–human will, even nature–history. Nature is what humans are not or what humans do not produce, what exists and "acts" without human influence, i.e., what is from itself, arises spontaneously. In conservation concepts this has been expressed its strongest form – but by far not only – in the narrative of "Humans apart from nature" and, linked to it, in most expressions of the wilderness concept.

Especially this type of dichotomies has played a major part in many philosophical as well as conservation controversies and has been deplored as a dichotomy with highly negative impacts on our relationship with non-human nature. But beware, the issue is far from simple and that on various grounds. In philosophy and anthropology there has been a long tension or even dialectics as *to what degree* humans are part of nature (due to their bodily existence and as being subjected to the laws of "nature") or *to what degree* they transcend nature, especially by means of their uniqueness or particularity, as possessing reason (or as being created in God's image). These discussions are about "human nature" (the essence of being human), which also links human societies to nature. It raises questions such as should we follow nature or should we transcend nature? Depending on the specific philosophy, the answer may swing in both directions, but overall, humans are (shaped by) both, nature *and* culture (Soper 1995).

What is natural? Naturalness as a descriptive and a normative concept

The term "nature", beyond the understanding of nature as everything that exists, is not merely descriptive; it is also normative or, more correctly, used in a normative manner. Various values are attributed to it, often more or less unconsciously. Nature and the natural are often used to justify political actions, especially in conservation, with the "natural" being seen as the "good" thing, which people and society must strive for. In contrast to most assumptions in conservation, however, nature is not always seen as positive only – or not at all. While some philosophers, such as Jean Jaques Rousseau, saw the (at least hypothetical) "natural state" of humans as the most pure and innocent, from which humans have fallen, other philosophical traditions postulate that the natural condition and natural desires have to be overcome to become really human, not the least by morality. Nature, here as the untamed part in humans, may not always be the best guide for keeping society together. The discrimination of certain human properties, behaviours, and preferences (e.g. race, sexual orientation) as "unnatural" has – and still is – often been used to oppress people, even though such judgements are in fact cultural ones. Likewise, non-human nature is not always experienced as beneficial but can be frightening, threatening, detrimental, and even lethal to humans, most dramatically shown by diseases such as the COVID-19 pandemic, by predatory animals, or by "natural disasters" such as earthquakes.

The different interpretations of the historical development of humans in relation to non-human nature form also an important (even though mostly implicit) part of the narratives according to the structure introduced in Chapter 3. The step from hunter-gatherer society to an agricultural one, e.g., is partly interpreted as progress and a step towards emancipation from the vagaries of nature (i.e. positive), but partly already seen as degeneration of previously good relations with nature.

So what is the "natural" in conservation, how do we assess it, and to what degree is it "good"?

The extreme positions sketched above are not very helpful to analyse ideas of the natural in a conservation context. If everything is nature ("nature" as everything that exists in the biophysical sense), also all human actions are natural, including those that most conservationists would call destructive to non-human nature, such as mining or the destruction of the Amazonian rain forest. If, on the other extreme, "nature" and "natural" are only what is completely unaffected by humans, there is no place on earth that is natural any more, as air pollution and the effects of human-made climate change are omnipresent in every corner of the globe. So if protecting only "natural" things in this sense would be the goal of conservation, there would be nothing left to protect. The ideas of the natural that most conservationists mean must therefore be located somewhere between these extremes. The "natural" must be conceptualised more as a *gradual* thing rather than a strict dichotomy. Thus a wooden cabin is more natural than a concrete building, a classical wilderness area is more natural than a designed urban park, and the park is more natural than a greenhouse (see also Rolston 1990 for a good discussion on this issue in the context of Yellowstone National Park). But is the most natural in this sense (least human influence, past and present) always the best? Are traditional cultural landscapes, in Europe, Asia, but also Australian native lands, just necessary compromises as compared to "pristine" ones, or do they have specific high conservation value? The devil is in the details, both when we try to asses to what degree something is natural and also in terms justifying the "natural" as the "imperative for biological conservation" (Angermeier 2000). Why should the "natural" be normative? Can it be taken for granted that striving for the least human influence is always the most desirable? After all, this position tends to reinforce the often-lamented separation between humans and nature.

Determining the naturalness of an object, an area, or a human activity is not straightforward and different approaches exist, which in conservation also vary depending on the specific site that is at stake.

In conservation "natural" is variously also equated with "intact" or "wild" (but see discussion by Ridder 2007). "Natural", or the degree of naturalness then, is either determined by an actual measure of human influence (or its absence, respectively) or by comparison of the current state of an area to a particular historical reference state which is presumed to mirror this absence of human influence, in terms of species composition and/or particular processes. In a much-cited paper, Anderson (1991), e.g., used three criteria to assess naturalness, namely (1) how much a system would change when current human activities would stop, (2) how much energy would be needed to keep an existing system in its current state (e.g. mowing, control of animal populations, or other maintenance efforts), and (3) how many of the native species existing in the area before it was settled are still there today.

These measures for naturalness, however, raise some important follow-up questions. First of all, there is disagreement if all human *activities* are considered "unnatural". For Angermeier (2000), e.g., human activities are natural if they do not use technology, while Hunter (1996) defines all human activities as "unnatural", not the least on behalf of the practical difficulty to draw clear lines between different types of human activities. This links closely to the second question, namely which historical *state* of an area to select as that which was still natural. In North America and Australia, some conservationists see the advent of Europeans as the point in time until which the areas were still in a natural state, thus assuming that also the activities of native people where still "natural". A major argument here has been that the influence of American Indians and Australian Aborigines on non-human nature was rather "negligible", an argument that meanwhile has been refuted by archaeological evidence (e.g. Flannery 1994, Mann 2005). Others see the reference state for a truly natural state of an area only given as that before the arrival of *any* humans on the scene. Including or excluding some groups of people as components of "natural areas" (e.g. in the "Indian wilderness"; see Section 3.3) also shows that the overall actant "humanity" is sometimes split up into different actants, in this case "modern" humans and "traditional" humans (the latter referring to indigenous people). This more differentiated view of human actants can be productive and even necessary for conservation (as discussed in Chapter 5), but it can also be problematic, when it amounts to naturalising some groups of people as "wild" and contrasting them to "civilised" people, as was done frequently in the early days of conservation and in imperialistic thinking (see Section 2.2).

It should be noted that the notion of naturalness in conservation is very frequently linked to an emphasis of the *nativeness* of the species of an area. Valuing nativeness is partly a cultural, partly a scientific issue (see Box 4.1).

The different reference states for defining the "natural" are similar and partly identically to what I described as different perceived states prior to crises in narratives of human–nature relationships in Chapter 3. They are also close to some of the supposed starting points of the Anthropocene (see Box 3.4).

So even if we are able to determine the natural state of an area in one or the other way, why should it be the "imperative" for conservation? Philosophically, no *direct* inference from an *is* (here the natural) to an *ought* can be made (the well-know "naturalistic fallacy"). There always has to be a value argument to justify why we should follow nature, should aim at the natural. Something is not simply "good" because it exists independently of humans. Infanticide among animals, even primates exists, is "natural", but no one would take this a guidance (or at least excuse) for human behaviour. In turn as Soper (1995, p. 18f) has argued, if we see every human activity, even picking berries or hiking in the woods, as something that "devalues" nature, why do we not apply this view also to other species, who also prey on each other? Why do these other species and their activities not devalue nature? Again, in putative favour of nature conservation, we tend to reinforce the dichotomy between humans and non-human nature.

In fact, not all conservationist agree that the aim of conservation should be to take back as much human influence as possible, as also the narratives and the major conservation approaches discussed in Chapter 3 show. Especially in Europe, cultural landscapes, seen as a joint product of non-human nature and humans, ideally in "harmony" between humans and nature, are a major conservation object. Several types of grasslands are typical examples of highly valued conservation areas, grasslands which were created by extensive agricultural activities, both in xeric and in wet areas. These grasslands are not only protected on behalf of their aesthetic character but also for maintaining rare plant species which were promoted by these agricultural activities, sometimes from other parts of Europe (i.e. they originally were "exotic"). The old agricultural practices have to be continued or at least mimicked to avoid the disappearance of these habitats through natural succession (see also the example of heathlands in Section 3.4). From a position where naturalness in the sense of least human influence and management is the benchmark, such landscapes would be considered as "unnatural" or even degraded. Likewise, many restoration activities do not return particular places to their "pristine" (natural) conditions but nevertheless create sites of value for conservation. So the degree of "naturalness" that we strive for as goal for conservation is a matter of societal choices, not the least depending on the history of an area. Conflicts, however, prevail, both on behalf of misunderstandings about conservationists' implicit (different) conceptions of nature and naturalness, as well on different opinions of what should guide conservation. They are not always easily resolved, even when the existing ideas of "nature" are made more explicit (see, e.g. the case study of the Oostvaarderplassen, Section 3.1).

BOX 4.1 NATIVE VS. EXOTIC SPECIES: ARE SOME SPECIES LESS NATURAL THAN OTHERS?

It is quite a common postulate that conservation should focus on *native* species and ecosystems, with the idea that *exotic* (or *alien* or *non-native*) species are detrimental to native ecosystems and that they compromise the naturalness of communities and ecosystems.

Defined rather neutrally, *exotic* species are species which occur outside of their native range; the occurrence of the species must have been prevented in the past by a barrier to dispersal, and not by the conditions in the new habitat

(Heger and Trepl 2003). An *invasive* species is then any species that occurs and *spreads* at a location outside its area of origin. These definitions do neither presume human agency in the process of dispersal nor "negative" impacts with respect to matters of biological conservation (but see other definitions, as discussed, e.g., by Heger and Trepl 2003). Beyond this neutral, merely descriptive use of terms, a sharp distinction between "native" and "exotic" species is often made, with native species (and native biodiversity) valued positively and exotic species valued negatively. This distinction even overrides the normally positive evaluation of high diversity. While the argument is often more about *invasiveness* of species than about the mere property of being non-native, e.g. Angermeier (1994) argues that non-native species ("artificial diversity") should even be excluded from the definition of biodiversity. Likewise, alien species have been called "biopollution" (e.g. Olenin et al. 2007) and thus also excluded from biodiversity proper. Thus it appears that exotic species are often considered as less "natural" than native ones.

The language and discussions on dealing with "exotic" and "invasive" species are sometimes highly emotional, not the least as they are connected to various metaphors that allude to (undesired) human immigration or even xenophobia and "nativism" (Eser 1999, Woods and Moriatry 2001, Simberloff 2012). Cultural issues thus play a crucial – even though often unconscious – role in considering if exotic species are less natural than native ones. The distinction between exotic and native is based on a historical perspective, i.e. if species have moved into new environments and ecological assemblages or not, but there is also an ecological rationale behind it. This is the assumption that the "original" communities are the result of species having a long history of mutual adaptation with each other and the abiotic environment, leading to a balanced and properly functioning system. These systems are then seen in danger of being degraded or destroyed by exotic species, a claim that is, however, discussed controversially (e.g. Davis et al. 2011, Simberloff et al. 2011).

In Europe, with its millennia-long history of species introduction by means of trade from Asia and North Africa, a distinction is often made between "neobiota" and "archaeobiota". The latter are defined as those exotic species which have been introduced to Europe before 1492, the former as those which arrived after the discovery of the Americas by Europeans, being part of a huge new wave of exotic species. Archaeobiota are considered to be "naturalised" meanwhile (see also discussions by Hettinger 2001, Heink and Jax 2014).

In principle, however, there is no reason why exotic species as such should be considered less "natural" than natives ones, at least not on a biological basis. They are living beings like any others, and it is mostly impossible to say if a species is exotic or not without knowing its ecological history. It is only by cultural distinctions (e.g. seeing a particular historical composition of ecological assemblages as the "natural" one) and by human valuation that a different degree of "naturalness" can be meaningfully stated for non-native species.

Why human–nature dichotomies are not necessarily a problem for conservation

Complaints about a flawed and detrimental human–nature dichotomy (or dualism) are popular and easily made. In short, the argument is that, at least since the Enlightenment, the perceived separation between humans and nature, mind and matter, has led to environmental destruction because it morally devalues nature and reduces it to a mere resource for humans (see, e.g. Durcarme and Couvet 2020). But the connection is not that simple. On epistemological grounds alone it may even be necessary to make the distinction between humans and nature, at least in the context of current Western thinking. How can we even talk about nature conservation and analyse its links to society when no distinction is made between human and non-human? Büscher and Fletcher (2020, p. 130) express it:

> [E]specially because nature and society are inherently interrelated do we need to distinguish between their different elements; only in this way can we *meaningfully* understand the relations that constitute their inter-relation. If everything is ultimately and *only* one and the same, then we can neither understand the whole nor the parts.

Beyond that, how can we discuss human accountability towards non-human beings without that distinction? As Soper (1995) puts it quite succinctly:

> Unless human beings are differentiated from other organic and inorganic forms of being, they can be made no more liable for the effects of their occupancy of the eco-system [sic!] than can any other species, and it would make no more sense to call upon them to desist from "destroying" nature than to call upon cats to stop killing birds.
>
> *(p. 160)*

It is thus crucial to note that distinguishing *conceptually* between humans and (non-human) nature does not automatically imply an indifferent or negative *moral* position towards non-human nature (its devaluation). A problem for conservation is thus not human–nature dichotomies per se but their interpretation as a moral position, i.e. confusing the distinctness of humans from other parts of nature with a presumed human superiority over them.

What can it mean to overcome the dualism between humans and nature?

As discussed, the analytic distinction between humans and nature can be necessary for speaking about nature and about our relation to nature. It does, however, not automatically imply humans being understood as exceptional and superior. It is this status, the value we give to ourselves and the other parts of nature, that makes the difference and causes the problems, when seeing non-human nature as inferior and as a mere resource for human needs and desires. There is a close connection here between our self-image as humans, what it means to be human, and our understanding of and relation to nature.

Of course, humans are parts of nature in being an animal species, but of course they are also special, possess particularity – as does each other species. One specificity in which humans, as far as we know, are unique is our ability to take responsibility for our actions, which is the very basis why we – as many think – should protect nature. But what could it really mean to abandon the dualistic views of nature and culture? I have struggled and still struggle with this, because it is mostly trivial ("everything is nature") and/or of little practical relevance for conservation, being more a romantic, idealised view on nature. I think that most of traditional Western philosophy is not that helpful in this regard. But beyond the potential of non-Western philosophies (Section 4.3), especially anthropology, and possibly other disciplines that focus also more on human *practices*, may provide some understanding of what it might mean to move beyond human–nature or culture–nature dichotomies.

Tim Ingold, building on his studies of various hunter-gatherer societies, asks for rethinking our own relations to nature under the perspective of what he calls an "ontology of dwelling". The main point he makes is that we should take "the human condition to be that of a being immersed from the start, like other creatures, in an active, practical and perceptual engagement with constituents of the dwelt-in world" (Ingold 2000, p. 42). For Ingold, this is not another *world view* than our traditional Western one but another way of *apprehending* the world. In his words again:

> [A]pprehending the world is not a matter of construction [as in the Western approach] but of engagement, not of building but of dwelling, not of making a view *of* the world but of taking up a view *in* it.
>
> *(ibid.)*

Such a way of approaching nature and the human embeddedness in it is not unique to hunter-gatherer societies but has its reverberations or analogies also in other ideas such as feminist environmental ethics or the notion of O'Neill and colleagues (2008) as "living from the world, living in the world, living with the world". Similarly Berghöfer et al. (2010, 2022) operationalised the concept of "Societal Relationships with Nature" (SRN) not only via a cognitive dimension of knowledge and abstraction but also by two other dimensions, namely interactions with nature (material as well as non-material) and identity. No particular idea of what exactly "nature" is and no dichotomy between nature and culture is presupposed, and indeed in interviews conducted in southern Chile for developing the original SRN framework, some respondents had problems with understanding what was actually meant by "nature" at all.

Efforts to overcome the dualism between humans and nature in this way are still not the mainstream of conservation and conservation biology, where, in practice, the "natural" is still taken for granted, even though in multiple, mostly implicit meanings as described earlier. However, some of the narratives and concepts described in Chapter 3 have potential to develop this further, such as "Being at home in nature" and some expressions of cultural landscape approaches related to it, but also some varieties of concepts from the "Ecosystem management" cluster. Hopefully, the increasing focus on indigenous knowledge and co-construction of conservation with people from other cultures can bring some more changes in approaches to nature and naturalness in conservation (see Sections 4.3–4.5), not the least in the narratives underlying conservation concepts and the relative importance they have in conservation.

What do the natural sciences contribute to conceptualising nature? How is nature?

There is not one "nature" but different "natures" in the plural, depending on different philosophical ideas, different perceptions of the world, and different self-understandings of humans; this also became visible in the different major narratives in Section 3.3. Is "nature" therefore a mere human construct? This at least has been postulated by those anthropologists and sociologists who are leaning strongly towards culturalist theories of the world (see Descola 2013 for a critical discussion). I am convinced that it is not, as I will discuss in some detail in Section 4.2. But then, what role do the *natural* sciences play in conceptualising and understanding nature and naturalness? In fact the natural sciences have not much to say on *what* nature is but provide an indispensable contribution as to *how* nature is. Even constructivists have to acknowledge that there is a material basis to human life, which is beyond any construction, even if we may not grasp it in its every essence.

The issue of *how* nature is can again be considered from a descriptive as well as from a normative perspective, which are often interlinked. I have separated these perspectives in the general fabula structure for analysing narratives on human–nature relationships as developed in Section 3.2. Part of the normative characteristics attributed to nature have been touched upon already earlier in the current chapter. The question as to whether (non-human) nature as a whole or parts of it are considered to be good, bad, nourishing, sublime, frightening, hostile, etc. are in the eye of the beholder and shaped by cultural traditions and individual experiences. They always refer to particular relations and interactions between non-human nature and humans. The descriptive part is different, at least in part. It also can build on experiences, local and traditional knowledge of individuals and societies, but it also forms the very subject of the natural sciences, including ecology and parts of conservation biology.

Let me repeat in this place that I do not explore all ideas of nature and human–nature relationships shared among humans, but those relevant and prevalent in conservation concepts and practices. This is why I also have a special focus on ecologists and conservation biologists here. So what do they say about what and especially how nature is? First of all, they are of course normal people and share most ideas of what nature is with any other person in the Western world and beyond. But on behalf of their special interests, skills, and knowledge, they also have some special perspective on the subject of nature. In a project on "Visions of nature" Rob Lenders (2006) studied people's philosophies regarding nature in Germany, the Netherlands, and the UK. In this context he also explored the perspectives of ecologists – not on the basis of interviews with ecologists, but through an analysis of the literature. He distinguished and allocated their different "visions of nature" along two orthogonal axes or gradients: man-inclusive nature–man-exclusive nature, and nature in balance–nature in flux. The first is a philosophical gradient, as discussed earlier. The second gradient has indeed been a matter of much debate in the history of ecology: is nature stable, constant, or in balance (see Section 3.4, section on supportive concepts/the ecosystem functioning cluster) or is it ever-changing, frequently or always in disequilibrium and in flux? In the early days of ecology (and conservation), the idea of a balance of nature was dominating (Egerton 1973, Simberloff 2014). But at least since the second half of the 20th century an understanding of ecological systems as much more dynamic (and thus also less predictable) began to become the mainstream of ecology (see Pickett et al. 1992), which was also highlighted by conservationists (e.g. Wallington et al. 2005). In the context of conservation, other important properties either of nature as a whole or of parts – such as populations, communities, and ecosystems, even social-ecological systems – have been discussed. This refers to the question as to whether nature (or its parts) are robust – or at least resilient, – or if they are fragile and vulnerable, especially with respect to human influences. In principle, the natural sciences should be able to answer these questions. However, they cannot do it for "nature in general". As to whether something is robust or fragile, resilient or not, depends on many specifications. Nature as a whole (in terms of the biosphere) may be very robust when it comes to exist as a biosphere at all (a continuity of life on earth), but the answer is different when it comes to the maintenance of specific appearances of the biosphere, such as containing specific species – e.g. whales, humans – or specific ecosystems (such as current polar ones). The same holds for ecosystems: does it need a particular species composition that has to be maintained to still speak of the "same" ecosystem (see Jax et al. 1998) and to judge it as robust or stable, or a specific suite of processes, or a particular physiognomy (forest but not savanna or grassland), or even the continued delivery of ecosystem services (Jax 2010)? A general answer – which characteristics "nature" or ecosystems overall possess – is thus difficult to obtain, if at all, and depends also much on the type and location of a system. That means, much of it is a matter of empirical scientific research in specific settings.

In terms of conservation, however, applying the precautionary principle and thus not simply *assuming* robustness and resilience without further specification and scientific knowledge, is advisable. Such a principle also often appears behind the choice of "natural" conditions as a benchmark for conservation and restoration: we know – or at least think that we know – that nature worked well without humans in the past ("nature knows best"). But the world is complex and changing, even without human interference, and the "natural imperative" (in the sense of nature being best off without human influence) is certainly not the solution to all conservation challenges, especially when brought forth in a puristic manner (see Chapter 5).

An intermediate conclusion

To sum up, there is not one "nature". Ideas of what nature is have changed since antiquity in Western thinking, with several different notions of what is nature existing concurrently all the time. Nature and the natural were mostly defined in juxtaposition or even opposition (i.e. as a dichotomy) to the human realm, namely that of culture or society. At the same time, as e.g. Haila (2000) has noted, nature also served positively to define human self-understanding and identity. This can be seen in conservation narratives such as those of "Being at home in nature" and in many concepts of a good and truly human life (for the latter see Sections 3.1 and 5.3). The border between humans and nature has been questioned and re-thought all the time, as emphasised by Soper (1995, p. 73). While dichotomies between humans and nature have had some problematic practical influence on Western handling of non-human nature and also on other humans, today it might be that, as Kopnina (2020, p. 6) writes, the problem is not any more "whether or not humans are part of nature – of course, in evolutionary terms they are – but whether their influence endangers and discriminates against other species". The reason is that distinguishing humans and (non-human) nature analytically is not necessarily a value judgement about human exceptionalism, superiority, or the right to dominate nature and using it as a mere resource.

The notion of the "natural" (in all its various facets) as a benchmark and guidance for conservation and restoration is still highly prevalent. While useful in many cases, it is laden with both the difficulty of assessing naturalness and with the question to what degree naturalness is really an appropriate guidance for many conservation contexts. The actual (and potential) value of both well-established cultural landscapes as well as new hybrid, novel ecosystems should not simply be discarded at first sight because they are not as "natural" as "pure" and "pristine" wilderness areas. Naturalness, in particular expressions, can also be misguiding and counterproductive for conserving specific objects of conservation in the face of changing climate and, in addition, with respect to bringing together the needs of human populations and non-human nature.

It is not necessary to always start with a fixed and clear definition of "nature" in a conservation context (see Berghöfer et al. 2022): *The very idea of what and how is nature* is partly developed within the process of exploring one's idea of our role(s) in relation to the non-human world – or in analysing it within conservation concepts. But there should be awareness of the various understandings of nature and the natural, and the specific understanding(s) at hand should be made explicit. Developing an idea of nature is not just an academic "exercise" but gives way to action. Nevertheless, different ideas of "nature" and of (proper) human–nature relationships may clash. But it is then important to identify this as one source of conflict, especially if these ideas have been implicit up to then.

This perspective, that there is not one "nature", as something fixed and as clearly given, has led to heated discussions in conservation, especially with proponents of the wilderness idea. Fears were expressed that such a "relativistic" approach towards nature would be highly detrimental to conservation. This debate will be explored in Section 4.2.

4.2 *Case study:* "Does nature exist?" The debate on constructivism and its implications for conservation[2]

In 1995 two edited volumes were published whose title or subtitle, respectively, was equally "reinventing nature" (Soulé and Lease,1995, Cronon 1995). The timing and the titles were no mere coincidence. The background of both books was a project with the very name initiated by the University of California's Humanities Research Institute at Irvine. Between 1992 and 1993, three conferences were organised at different places in California and finally a larger interdisciplinary group of scholars, led by environmental historian William Cronon, worked and lived together at the Irvine campus for a whole semester in 1994. Their explicit task was, inter alia, to produce a book, as one result of their work. Conservation biologist Michael Soulé and philosopher and historian Gary Lease (both not part of the Cronon group) were both intrigued by the subject of the project and somewhat unsatisfied by the content of the conferences. They felt "that none of these conferences was designed to address specifically the dialogue between the worlds of the sciences and the humanities" (Lease 1995, p. 7). In consequence, they organised an additional two-day conference to fill this perceived gap in 1994, also supported by the "Reinventing nature" project. Their book was an outcome from this latest conference.

Although the two books, as they were produced almost simultaneously, do not exactly "respond" to each other, they illustrate two sides of a debate about the "reality" of nature. This debate had hitherto taken place mostly in the humanities, namely under the heading of constructivism, meaning the idea that what we consider as "nature" is not given to us as such, and simply by our senses, but that it is the results of cultural and societal "constructions". The two books, and the further reactions to them, demonstrate in which way the debate is relevant to conservation and conservation biology.

Soulé and Lease, and most authors of their book, fiercely attacked any constructivist approaches as detrimental to conservation – as already the question mark behind "Reinventing nature?" and the subtitle of the book ("Responses to postmodern deconstruction") indicate. Beyond their famous statement that the authors of the book "agree that certain contemporary forms of intellectual and social relativism can be just as destructive to nature as bulldozers and chain saws" (Soulé and Lease 1995, p. xvi), they further speak, in militaristic parlance, of a "war over nature [being] in progress" (Lease 1995, p. 4) and about a "siege of nature" (Soulé 1995) which such constructivist views were bringing about. In the words of Soulé:

> Why this book? Gary Lease and I were concerned that the wave of relativistic anthropocentrism now sweeping the humanities and the social sciences might have consequences for how policy makers and technocrats view and manage the remnants of biodiversity and the remaining fragments of wilderness.
>
> *(p. 159)*

Cronon's book *Common Ground*, on the other side, being more open towards facets of constructivist thinking, had at first indeed used the project name "Reinventing nature" quite affirmatively in its subtitle (1995). But already from the second printing in 1996 on the subtitle changed it into "Rethinking the human place in nature". I did not find an explicit mentioning why this change took place, but one possible reason could be the harsh reactions the book received, accusing the authors of promoting and abetting anti-environmentalist positions. In the foreword of the paperback edition (1996), Cronon thus felt obliged to emphasise again that "the last thing we want is to undermine the [environmental] movement or its long-term agendas" (p. 20), mirroring a similar statement already in the original introduction. Cronon states that the main goal of the book's criticism on traditional ideas of nature was "to encourage greater reflection about the complicated and contradictory ways in which modern human beings conceive of their place in nature" (ibid.).

So what is at stake, what does a constructivist perspective towards nature mean and what could it imply for conservation? "Social constructivism" can coarsely be portrayed as the position that the answer to the question of *what nature is*, is not given by nature as such but that nature is a social and cultural construction. There is thus not one nature, there are many different *natures*, depending on cultures and perspectives. If, however, nature and parts of it (communities, ecosystems), also "wilderness", can meaningfully be defined in many different ways, this may be interpreted as a *relativistic* perspective towards nature, where there is no certainty (e.g. through the natural sciences) about what and how nature is. But is nature really completely relative, is it merely a construct, or many different constructs – even a figment of human minds? What then could protecting nature mean at all? Does it mean that "anything goes" with respect to resource and ecosystem management – an accusation that has been made, e.g., against what Cronon's book would imply? The answer to these questions is crucial to what we see as worth conserving and what not, and how we select and justify our benchmarks for protecting and restoring nature.

The question as to whether nature (or ecosystems or wilderness) are constructs cannot be answered simply by either "yes" or "no". In fact, both are correct. It needs a closer look on a broader, sometimes very heated controversy during which many strawmen have been erected and burnt. Criticism focuses on three major questions: first a philosophical one: what can we know about nature and what could it mean that nature is constructed?; then a strategic one: what does it imply for conservation if we accept that nature is – at least *also* – constructed?; and third: in which way do constructivist approaches to nature reinforce the lamented human–nature (or culture–nature) dichotomy and thus cause additional problems for conservation and the human–nature relationships linked to it?

What can it mean to speak of nature as "constructed"?

The position of constructivism, or more specifically the claim that particular things, such as gender, emotions, danger, quarks, wilderness, ecosystems, and even nature as a whole are socially constructed (see, e.g. Hacking 1999, p. 1 for a list of things that are said to be socially constructed) has aroused many controversial discussions, both for philosophical and political reasons. The issue behind these controversies is of considerable importance for conceptualising and operationalising nature and for determining the goals of nature conservation. In particular, it can cause major misunderstanding if not dealt with in a differentiated manner. In order to stay focused, I will confine my treatment of this debate largely to discussions that refer to the realm of ecology and conservation biology.

So what may it mean to say that nature or wilderness are socially constructed? The extreme positions towards what nature "really" is and how we can talk about it are those of a *naive realism* on the one hand and that of a *strong constructivism* on the other. Proponents of a naive realism believe that we are able to directly perceive of nature through our senses as it really is and that this reality ("the facts") is independent of human perspectives, interests, and values. In contrast to that, a strong constructivism negates the existence of facts and nature as such and postulates that nature is a purely cultural phenomenon, depending on specific cultural perspectives on the world. Most philosophers of science reject naive realism, claiming that our perceptions of nature (or reality in general) are always filtered and influenced by our previous cultural backgrounds and experiences (knowledge, theoretical assumptions, values, etc.) and by our interests. But there is also strong criticism of the constructivist position, from philosophers but even more from the camps of conservation biologists and conservationists. Part of this critique is political and strategical: if nature is just a human construct, there is, they state, no reason to protect any *specific* state of the world ("the" natural one). More than that, the discussion about our perceptions and modes of construction of nature is also said to distract us from solving the ecological crisis or even would question if there is a crisis at all (Soulé and Lease 1995, Kidner 2000, Crist 2004). Another part of the critique is philosophical: how is it possible to negate the very existence of a nature as existing independently of us and our ideas, interests, and attitudes?

In his book *Pandora's Hope*, the French sociologist Bruno Latour, being one of the most cited "constructivists", recounts a meeting with another scientist (a psychologist) who had read his writings. "Do you believe in reality?", he was asked. Latour, somewhat surprised, amused, and embarrassed at the same time, replied: "'But of course!' I laughed. 'What a question! Is reality something we have to believe in?'" (Latour 1999, p. 1). This little episode displays quite well the kinds of misunderstandings that exist with respect to what it means that nature is socially constructed. As several authors (Hacking, 1999; Proctor 1998, Demeritt 2002) have demonstrated, the statement that something (nature, wilderness, and other things) is socially constructed – or the accusation that someone considers something as "just constructed" – is prone to misinterpretation, because different things are meant by "socially constructed".

First of all, "constructed" may refer to different things. Ian Hacking (1999) already in the title of his book thus asks: "The social construction of what?" The "what" may either be an object, a concept, a classification. He and others agree that the most important distinction here is between the construction of a concept and that of an object. This also alludes to a distinction between epistemology and ontology. Do we mean that a *concept* of something is/has been socially constructed or is the *object* as such constructed and does not exist without humans? Do we mean, as two ways to understand the latter, that we have constructed the object physically, by physical human actions, or by our mental acts? There are considerable differences between different groups of "constructivists" with respect to these questions (Hacking 1999, Demeritt 2002). While a few constructivists would really argue that language and social processes are the only "real" things, preceding any physical reality, in general, most constructivists, as the quote from Latour demonstrates, do not deny the existence of some kind of reality existing independently of us, even though they strongly question how much we can really know about it in terms of "objective" knowledge.

Also Cronon, while strongly emphasising the cultural dependency of concepts of nature, explicitly states in the introductory essay of "Uncommon Ground":

> The reality of nature is undeniable. The difficulty of capturing it with words – not even with the word "nature" itself – is in fact one of the most compelling proofs of its autonomy.
>
> *(Cronon 1996, p. 52)*

There is often a close relation between conceptual and physical construction of nature (and other things). Humans of course have always modified and thus constructed nature (here the physical world) and at the same time imagined (constructed) what nature is (to them). As Demeritt (2002, p. 779) emphasises:

> Clearly many material constructions of nature will depend on the conceptual constructions that guide the ways people interact with and transform the physical environment, which in turn will influence what people conceive.

Furthermore, there are different categories of what "nature" means to those who argue about its construction. The debate is in fact not about the existence or non-existence of nature in the more generic and extreme meaning of nature as everything that exists, physical nature, the "world outside" (Demeritt 2002, p. 778, see also Rolston 1990). It is (mostly) a strawman erected by opponents of constructivism (or "deconstructivism", as they sometimes call it) that this kind of nature is seen as denied or is considered as being constructed by constructivist philosophies. Most of the proponents from the constructivist side argue about specific expressions of the "nature of nature" (i.e. what it means of something to be "natural") and our (in) ability to describe or recognise it with certainty. In fact, the discussion is one about epistemology, about our abilities to know about nature and not so much about ontology, i.e. as to whether nature exists and what it "really" is (Proctor 1998). Or, as Philip Kitcher put it succinctly:

> So do we construct the world? In the sense often intended in fashionable discussions, we do not. There is all the difference between organizing nature in thought and speech, and making reality: as I suggested earlier, we should not confuse the possibility of constructing representations with that of constructing the world.
>
> *(Kitcher 2001, p. 51)*

But even such more moderate (epistemological) claims are attacked fervently and also often mixed up with an ontological perspective:

> Such [constructivist] claims suggest that nature is an entity very different from that which many environmental theorists, writers, and activists have up to now believed. Rather than being viewed as a multifaceted, diverse order whose pattern and possibilities extend well beyond our abilities to understand them, nature becomes an offshoot of *social* reality which also constructs individuality. And since the social world varies according to time and place, then it follows that each of these social worlds will construct a somewhat different version of nature, and there is, therefore, no single "nature", but rather a "diversity of natures", constructed by our various fantasies and languages.
>
> *(Kidner 2000, p. 340)*

Also Soulé (1995), though acknowledging the existence of different "constructions of living nature" or "cognitive formations", is at pains to establish or maintain the priority of scientific (especially biological) knowledge above all other approaches, as the only solid ground towards conceptualising and understanding nature. This often goes along with accusations by conservationists (also by Soulé) that constructivists are ignorant of biology and ecology (e.g. Foreman 1996/1997, p. 4, arguing against Cronon).

Without relating this explicitly to the philosophical debates described in this section, debates in conservation are led – mostly from a realist position – not only about epistemology but about the ontology of nature: what "real" or "true" nature is. Soulé, e.g., writes:

> At the physical level, constructivists claim that nature is no longer natural. Rather, it is a human artifact because aboriginal peoples have altered it fundamentally with their economic manipulations.
>
> *(Soulé 1995, p. 148)*

This strong statement is valid, however, only if one would embrace an extreme and puristic ideal of "nature" as that which is completely "untouched" by humans, which neither Cronon nor most other conservation-minded "constructivists" do.

Neither naive realism nor extreme constructivism are really satisfying when it comes to discussing nature and the natural. There are thus many authors which, in different ways, argue for a middle position between these extremes (e.g. Pickering 1995, Soper 1995, Proctor 1998, Kitcher 2001, Descola 2013). My own position also dwells within this middle ground. Although I acknowledge the issues posited by a constructivist view as meaningful and important with respect to nature, I prefer to express it in other words. Of course, nature (as the physical world) is "out there" (in an ontological sense). The way we can perceive of nature, however, what we can know about it (the epistemological dimension) is clearly in our minds. It is influenced by our cultural backgrounds (in the broadest sense of the term) and by our specific interests. In other words, nature is *also* socially constructed. It is a concept that describes our relation to the world. Nature is a concept originating from human minds but at the same time it is "something out there", to which our concepts refer, and which the sciences, e.g. ecology, describe, as well as lyrics and painters do. This "something out there" is largely independent of us ("autonomous"). We cannot know nature in totality (its essence, or what it "really" is – the "things as such", as Kant expressed it), but nevertheless physical nature and its parts resist any purely arbitrary definitions. All the conceptual constructs we make of it, when applied to a specific situation, have to stand a test, namely if they fit the situation, e.g., in terms of predicting the dynamics of the things we refer to in our concepts, or even managing them. We may construe a concept of a stone wall as an easily permeable boundary between two places, but as soon as we test this definition by trying to move through the wall, the resulting headache suggests otherwise.

Usefulness in a practical, pragmatic sense is one test which concepts should be able to pass. As Pickering (1995, p. 22) put it: in the course of practice, nature *resists*, forcing us to accommodate the concepts we construct about it. Several authors have pointed out, that word "object" – or likewise "*Gegenstand*" in German – from its etymological roots contains the notion of resistance: "to object", to "stand against". A similar idea was formulated by Elizabeth Bird, saying that we are, in doing science, negotiating not only about the interpretation of our observations but also about reality. As she expressed it:

> [R]eality is being negotiated at the same time as its theoretical construction. And both of those, the reality and the interpretation, are not merely social constructions, but at both levels negotiations with nature. Nature's role in that negotiation takes the form of actively creating something materially new and resisting or accommodating the range of metaphorical and theoretical imaginings with which it is approached.
>
> *(Bird 1987, p. 259)*

Does the diversity of concepts of nature promote environmental relativism?

Constructivism is a highly emotive issue in conservation, as can be seen by the language through which it is contested. Thus, the environmental magazine "Wild Earth" devoted almost a complete issue (Volume 6/4, Winter 1996/1997) to challenge Cronon's book and opinions, partly in a biting tone, with allegations and personal attacks on Cronon. The most ardent critique of constructivism does not come from a philosophical perspective but from a political one. It arises from the fear that, if "nature" is not clearly fixed and predetermined, we end up in complete relativity. Relativity means that we may loose firm ground with respect to our perceptions and our knowledge about how the world "really" is, undermining the credibility of science and scientists as those who describe reality. In addition, it refers to moral, and thus political, relativity – summarised here for the sake of simplicity as "environmental relativism". The latter amounts to the allegation that environmental problems are not taken seriously, that they are just described as attitudes instead of facts, and that they thus further anti-environmental tendencies ("Why worry when climate change and biodiversity loss are just social constructs?"). Even more than the epistemological problems described earlier, this issue has raised much concern and even rage against constructivists' accounts of nature. The problem has been discussed in several places and with some polemics. Opposition has not only been expressed to *sociological* perspectives on the matter, but also to "relativistic" *ecological* concepts and theories. Thus environmental ethicist Baird Callicott complained about what he called "deconstructive ecology" (Callicott 1996). By this he meant the change from the notion of equilibrium ecology to non-equilibrium ecology, which took place in the 1980s. He argued that this new ecology undermined the reference states that conservationists envisioned for ecosystems and thus was also in danger of undermining the scientific basis of some concepts of environmental ethics, in particular that of Aldo Leopold's *Land Ethic*, the latter postulating that "a thing is right when it tends to preserve the integrity, *stability* and beauty of the biotic community" (Leopold 1949, p. 224f, my emphasis). Callicott, quite surprisingly, also subsumed Michael Soulé among the "ecological deconstructivists" because Soulé, in his book contribution in 1995, refrained, in a very differentiated manner, from the idea of nature as being stable and in equilibrium, which he (Soulé) decried as being outdated and a myth.

An argument along the same lines was brought forth in the mid-1980s by a prominent German ecologist, at that time minister for the environment of a German state. During a discussion about the goals of conservation in Germany, he argued that any deviation from the received equilibrium view of ecosystems (or more broadly, the "balance of nature") would lead to confusion and a loss of confidence in ecologists among the population. As a consequence, following the further reasoning, this might lead to decreasing public support for biological conservation and environmental protection policies. In a similar sense, arguments were put forth in American conservation biology. Beyond the question as to whether the notion of the balance of nature is necessary in arguing for conservation, Soulé and Lease (1995) were convinced that, in general, questioning the concepts of nature and wilderness was "sometimes [used] in order to justify further exploitative tinkering with what little remains of wilderness" (Soulé and Lease 1995, p. xv; similar and even more fierce arguments were raised in the issue of "Wild Earth" mentioned).

These are grave accusations. Are they justified? In a certain manner, the way we define nature is really relative. But that does not mean that there is a complete idiosyncrasy in the sense of a specific definition by each author or for each single empirical case. Nor do the various manners in which nature is conceptualised render environmental problems as purely relative, as some authors fear.

In terms of dealing with environmental problems, the idea that nature and ecosystems are partly social constructions does not, in fact, imply *moral* relativity, as many authors ascribe to constructivist approaches (e.g. Kidner 2000).

Environmental problems and conservation problems are human problems, not problems of nature. The phenomena within nature (e.g. climate change or species extinctions) that we *perceive* as problems are real, but it is only via human judgements that they appear as *problems*, that they *become* problems. In this way environmental problems are, by necessity, constructed, both physically – through human impacts on the environment – and conceptually. But it would, as several authors (e.g. Gandy 1996) have pointed out, be logical nonsense to insist that environmental problems are *mere constructs* (also in an ontological sense) and exclusively exist as matters of language and discourse. Otherwise the ozone hole, climate change, or biodiversity loss would physically disappear once we change the discourse about it, without any further physical action needed.

What we observe and judge as environmental problems are the consequences of changes brought about by humans to themselves or to other humans *via* the environment: effects on our health, economy, our aesthetic and moral sentiments, and effects on the parts of the non-human world which we need and/or value. And it is just these affected *needs and values* that make changes in the environment an environmental problem, regardless of the "natural state" of ecosystems, which is, as we have seen, at least difficult, often even inapt as guidance. But that does not diminish in any way the urgency of problems.

Not all ideas about nature are on equal ground. Material reality, as also, e.g., Cronon (1996, p. 22) emphasises, constrains our ideas and actions. But nature alone, as discussed before, even if we could unambiguously and exactly know what and how it is, cannot tell us how to protect it and what the best human relationship to it is. We need choices for that, sorting out between the desirable and the undesirable. Also these choices are not at will, there are rules and constraints, partly political and social, partly biophysical ones, and partly those of a consistency of moral arguments. This is where ethical considerations can support conservation. *Value judgements* are necessary to decide as to whether a specific patch of the world can be considered to be "intact nature" or natural in some degree, and if and how we might want to protect it.

Constructivism and the human–nature dualism

A third critique against constructivist approaches to nature brings back the issue of the human–nature dualism discussed in Section 4.1. Some critics of constructivist approaches to nature argue that these approaches – often contrary to their own intentions – were perpetuating this very dualism.

Peterson (2001, p. 73) writes:

> In principle, social constructionist arguments hint at the possibility of deconstructing boundaries between humans and "a genuine Other" – nonhuman nature. In practice, however, most constructionists shrink from that potential and instead enshrine discourse as something only humans can do. This reinforces a perception of humans "as ontologically privileged beings, set apart from, or even against, the rest of nature".

In consequence she argues for a "chastened" or "constrained" constructivism. What it amounts to is, again, the question of our own self-understanding as humans in relation to nature. But as discussed above, much depends on how we understand this dualism, as a question of conceptual distinction or as an ontological one with moral implications.

To sum up, nature is neither there "as such", universal and open to us to read in an unambiguous manner, nor is it a mere construction. It is there, as a physical object, but how we conceptualise it is still a matter of our construction. We need to find a middle ground here, allowing for various narratives but also without seeing every thinkable narrative as equally valid. We also should be careful when it comes to deriving moral statements from the partial relativity of different possible "natures". The problem that Soulé and others describe in fact exists and persists, namely that arguments about different constructions of nature, such as those discussed by Cronon and colleagues, receive applause from the wrong side, are used as an excuse for carelessness and further unrestrained exploitation of nature. Even when scholars speak clearly against this misuse and elaborate their arguments in a very differentiated manner, an overly simplified version of a discourse about different ideas and constructions of nature (or wilderness) is easily at hand and used by those whose interests are, in fact, against nature conservation. But this must be no reason to avoid such discussions. A reflection of concepts of nature and naturalness, of the cultural dimensions of nature concepts and embeddedness of humans within nature, as presented in Cronon's book, is essential (and highly inspiring) if we do not want to lead dogmatic discourses about the one and only "right" direction of conservation, discourses which can weaken conservation efforts much more than open and balanced debates about "the nature of nature". If they are not conducted by us, within conservation and conservation biology, they will be forced upon us by people really hostile to conservation and environmental issues.

4.3 Ideas of nature and human–nature relationships beyond "Western" thought

If nature conservation is not to remain a purely Western enterprise, today dominated by science, it also has to build bridges to "non-Western" and indigenous cultures and their ideas both of nature and of human relationships to it (including that of being "one with nature"). Building such bridges is, however, far from easy.

There is a great diversity of ways in which nature is conceptualised in non-Western cultures, even if one focuses at first on the linguistic level (Coscieme et al. 2020, Ducarme et al. 2020, Droz et al. 2022). The subject of ideas of and relations with nature beyond the Western mainstream of thought has been approached from several perspectives, in particular from environmental ethics (Callicott 1994), from anthropology and cultural studies (e.g. Descola and Pállson 1996, Selin 2003a), and by far not the least from the study of traditional ecological knowledge and practice (Berkes 2018).

More than in any other part of the book, I am here reaching beyond my core expertise. Given the complexity of the subject matter and the diversity of cultures, my cautious rapprochement to the theme can therefore in no way be comprehensive and will be limited to the analysis of selected studies and reviews. It might, however, at least describe directions for moving on further.

In the following, at first some caveats will be emphasised, which any exploration of non-Western human–nature relationships faces. Despite the high diversity of cultures and their relations to nature, there appear to be some similarities between different traditional and indigenous approaches regarding these relations, which will then be described. Both commonalities and differences will then be exemplified by short descriptions of some well-documented traditional approaches of relating to and dealing with nature.

Before continuing, a few words about the terminology in the field of traditional ecological knowledge are in order. Terms around are indigenous, traditional, and local knowledge, as well as practices and world views of the same, often intimately related. Specifically for *ecological* knowledge Berkes (2018, p. 8) defines "traditional ecological knowledge" (TEK) as:

> a cumulative body of knowledge, practice, and belief, evolving by adaptive processes and handed down through generations by cultural transmission, about the relationships of living beings (including humans) with one another and their environment.

"Traditional knowledge" may be seen as an overarching term. It is not restricted to indigenous people. Likewise, "local knowledge" may also be the non-scientific, experiential knowledge developed and transmitted by non-indigenous rural communities (see also UNESCO 2013, p. 68f).

Caveats

An ongoing discussion has been as to whether non-Western people, especially those of indigenous communities, are or have been "noble savages", treading the earth only lightly and being kind of born environmentalists or conservationists. The counter position is that these people in fact have been, as Berkes (2018) called it, "intruding wastrels". This means that people in general, also indigenous societies, were destroying pristine ecosystems, e.g. by the "Pleistocene overkill" of large mammals during early phases of settlement, in particular in North America and Australia, or through highly unsustainable land uses leading to deforestation (see discussions in Kalland 2003, Berkes 2018). There is evidence to support both positions but in their generality both viewpoints on indigenous people are meanwhile characterised as "myths": the noble savage as a romanticising of indigenous people and the "natural state" of humans, and the "intruding wastrel" as an echo of old colonial ideas of uncivilised "primitives". A third myth is described by Berkes 2018 under the name of "noble savage/fallen angel", by which he describes Western expectations of indigenous people as noble savages which then, "spoiled" by Western culture, were lead to a deviation from their "born conservationist" behaviour. In essence, for those following this latter idea, it means that traditional people "should continue to live as 'primitives,' lest they become a threat to the very ecosystems in which they live" (Berkes 2018, p. 250).

To avoid such oversimplifications, some caveats are in order (see Selin 2003b, Booth 2003). The first and obvious one is that there are always methodological problems in understanding different cultures by using the analytical tools of one's own culture. We already know how difficult it is to sometimes even provide proper translations between Western languages. So this of course is even tremendously more difficult when moving to different languages and cultures with completely different world views and mindsets. There is, what anthropologists are highly aware of, always the danger of – at least unconsciously – imposing one's own assumptions, framings, and concepts in interpreting ideas and practices of other cultures. Many Western conceptual categories, starting with "nature" and "culture" but also "native" or "alien" or even that of "humanity", are anything but universal human categories but instead are highly dependent on specific world views and epistemologies. For some (indigenous) people there is not even a word for "nature" (Zent 2015; see also later). In order to not presuppose a concept of "nature", Milton (1996), e.g., prefers to use the term "environment" as "that which surrounds". I will nevertheless continue to use the word "nature" in the following text but emphasise where local knowledge deviates from "the" Western concept of nature.

A second caveat concerns the question as to whether we look at historical ideas of traditional societies or at the way they are relevant for them today – affecting their daily lives and relationship with what we call nature. Traditional or indigenous societies are not static. Most of them have come into contact with and were influenced in various ways by Western societies with its economic mainstream and scientific world view. This does not mean that they have completely abandoned their old traditions, but substantial changes have often happened. Thus, traditional societies which were mainly hunters and gatherers, in addition, are today also using other means to meet their daily needs, e.g. through wage labour for modern companies (Milton 1996, p. 117). Also the recently popular Andean concept of *buen vivir* is a mix of traditional and modern ideas (Gudynas 2011; see case study in Section 4.4). There have even been cases where stories told by indigenous people on their human–nature relationship to researchers used the Western framings meanwhile expected by them, in order to gain more attention and support by Western organisations, such as conservation NGOs (Dove et al. 2003). This does, however, not mean that these stories are necessarily "invented".

In addition, as in Western societies, the ideals mirrored in narratives as well as diverse kinds of rules about proper human behaviours (between humans as with other parts of nature) do not always match daily practices. Kalland (2003) therefore asks for "humanizing the other", perceiving of indigenous people neither unquestioned as "noble savages" nor as (to use the expression of Berkes) "intruding wastrels".

A last caveat to mention here is that even if we avoid the mistake of lumping together all indigenous and traditional cultures, another danger exists when perceiving indigenous communities as internally homogenous. As described earlier (Section 2.6), also here the use of a common "we" (or "they") can be misleading. Also in traditional communities, differences in power exist and different actors sometimes have diverging interests (see, e.g. Löfmarck and Lidskog 2017). It is not always obvious who really speaks for the whole community or just for a part of it, or for his or her personal viewpoints and interests.

Some apparent commonalities of traditional and indigenous human–nature relationships

Even in Western conservation there is not one unifying, general human–nature relationship (see Chapter 3). Much less so in the highly diverse non-Western cultures. Nevertheless, some commonalities seem to exist, as several authors have emphasised. These commonalities – as much as in Western thinking – may not be completely universal but also not completely idiosyncratic. In terms of the patterns to be found in human–nature relationships (both Western and non-Western), Descola (1996) writes:

> But since patterns of relations are less diverse than the elements which they relate, it seems obvious to me that these schemes of praxis cannot be infinite in number. This is why I believe that the mental models which organise the social objectivation of non-humans can be treated as a finite set of cultural invariants, although they are definitely not reducible to cognitive universals.
>
> *(p. 86f)*

Berkes (2018, p. 18ff) describes four interrelated (nested) levels of traditional knowledge and management systems. At the first level he sees local knowledge of plants, animals, soils, and landscapes. Building on that (or surrounding it), the second levels relates to specific "resource management systems", including a set of practices. In order for such management systems to work, the third level contains social institutions, including rules, norms, and codified social relationships. The most embracing level then is the fourth, the specific world view, by which observations become meaningful and by which they are understood in a larger context of morals, belief systems, cosmologies (theories about the origin and order of the world). For all levels together, Berkes speaks of traditional knowledge as described by/as "knowledge-practice-belief systems".

This description already summarises some characteristics common to most indigenous human–nature relationships. I will briefly mention these and some others here and afterwards demonstrate what they mean in practice by four examples of traditional cultures. As the examples will show, not each of these characteristics is given in every non-Western culture but most of them are.

- First of all, for most traditional human–nature relationships, "nature", if the concept exists at all, is nothing abstract but closely related to personal and socially transmitted experiences of the non-human world, **knowledge and practice being intimately related**.
- Secondly, knowledge and practice are in most cases closely connected to particular **places**.
- Furthermore, **beliefs and religion**, mostly based on traditional cosmologies, play a crucial role for human behaviour with regard to non-human nature. In Western conservation, religion – i.e., Judeo-Christian religion (see Box 4.2) – mostly

plays no role, at least not explicitly; it sometimes still exists in an unconscious or secularised form. In most indigenous philosophies, nature – or at least parts of it – is sacred.
- Not the least on the basis of specific cosmologies, ideas of kinship and an **extended community of beings** are common in traditional societies. Such extended community can embrace other living beings but sometimes also spiritual beings (e.g. ghosts, ancestors, gods). These other "persons" are active actants, not just passive "nature".
- These other beings, as well as fellow humans, are seen as beings that demand **respect**, are often also in a relation of **reciprocity**, and are not just perceived as resources.
- Given that, for most indigenous peoples **caring for nature *and* using it** is no contradiction, but to the contrary, a necessary link, also in spiritual terms.
- Traditional knowledge, with all its moral implications as guiding (proper) behaviour, is often codified in the form of stories, i.e. in a **narrative form**.

There is another interesting aspect of the four levels by which Berkes describes traditional knowledge. Namely, I think that in principle these levels are not that much different when it comes to describing conservation and conservation science (and even "Western science" in general). Unless one is convinced that conservation is based solely on science and its immediate applications, conservation takes place in the context of social institutions and is also based on particular world views. Knowledge is both local and general, scientific and historical, and various social rules allow or constrain particular practices. Western science and the world view of the Enlightenment as well as – mostly more implicit – Judeo-Christian religion form the larger context. Differences are more a matter of degree than of kind, as has been discussed, e.g., by Agrawal (1995) or Turnbull (1997). For example, also indigenous knowledge is not devoid of generalisations nor is conservation science only based on generalisable globally valid knowledge. However, as concerns traditional and indigenous knowledge and related human–nature relationships specifically, the four levels described by Berkes are clearly much more intimately related than is the case for Western conservation and the emphasis in traditional human–nature relationships is clearly more place-based than much of conservation science.

The four examples following will show both the commonalities of different traditional human–nature relationships as well as their differences.

An indigenous world of peoples/humans: rainforest dwellers in South America and their cosmology

Cosmologies and religion play a major role in any understanding of humans and their role within nature. This often starts with creation myths, of which there are countless ones (see Leeming and Leeming 1994). In Western culture, creation is either Judeo-Christian creation or, as it comes to the conservation community, more commonly now an evolutionary cosmogony[3] which replaces – or at least appears to replace – the former. Common to both Western creation narratives – and to many others – is the idea that humans are the last and final part of creation, their appearance kind of completing it. So nature is first, then come humans, and, derived from human biology and behaviour, culture originates, which sets humans apart from the nature whence they came from. To a lot of indigenous peoples, not the least in the Amazonian basin, however, this sequence is completely the other way round. In their mythology, before the actual beginnings of time the world first consisted of humans only (Danowski and Viveiros de Castro 2017). But these humans were neither morally perfect nor a fixed entity; they had great bodily plasticity and could undergo multiple transformations. From this entirely human world, the whole world as we know it developed and, in a way, still remained essentially human (see below). As Weiss (1972) described for the Campa people of eastern Peru:

> When the curtain goes up, the actors are already on stage: the primal Campa, human beings living here on earth but immortal, many with powers exceeding those possessed by mankind today. [...] Whatever else existed in the universe at that time is indicated only sketchily in the mythology, but it was an impoverished universe lacking many features that would come into existence through transformation only with the passage of time. Campa mythology is largely the history of how, one by one, the primal Campa became irreversibly transformed into the first representatives of various species of animals and plants, as well as astronomical bodies or features of the terrain. In each case the mechanism of change was either the action of a transformer deity or auto-transformation.
>
> The development of the universe, then, has been primarily a process of diversification, with mankind as the primal substance out of which many if not all of the categories of beings and things in the universe arose, the Campa of today being the descendants of those ancestral Campa who escaped being transformed.
>
> *(Weiss 1972, p. 169f)*

According to Danowski and Viveiros de Castro (2017) this type of cosmology was not restricted to the Campa but is common in the Amazon. It can also be found in cultures beyond America, e.g. in Papua New Guinea.

In stark contrast to our common world view, thus not everything was "nature" first, part of it becoming human later, but it is just the other way round. As Viveiros de Castro (1998, p. 472) expresses it: "The original common condition of both humans and animals is not animality but rather humanity". Everything was human (and "culture", if you will) at first, and part of it became other species and what we see as non-living entities (stars, rocks, etc.).

This idea of creation has implications for the world view and the behaviour of Amazonian Indians up to this day. For these peoples, the whole universe is alive and forms a great network and a large assemblage of different communities. Other species, especially animals, also see themselves as people, i.e. as humans, with their own communities. Their form is highly mutable and transformations between different forms – also between humans and animals, and between animals or humans and spirits – occur. Viveiros de Castro speaks of "perspectivism" in this context:

> [A]nimals are people, or see themselves as persons. Such a notion is virtually always associated with the idea that the manifest form of each species is a mere envelope (a "clothing") which conceals an internal human form, usually only visible to the eyes of the particular species or to certain trans-specific beings such as shamans.
> *(Viveiros de Castro 1998, p. 470f)*

Being human, thus, is not a matter of the body but a property or a "condition" of beings. Decisive here is not being part of the human *species*, but it is the *personhood* of living beings, which makes them human – in their respective own internal perspective (see also Zent 2015; for the Sami people of northern Europe see Helander-Revall 2010).

So there is no sharp distinction between humans and nature. According to Zent (2015) there is not even a word or an expression for what we call "nature". There are clear distinctions between humans and other species, between "we" (the people) and "they". But also each other species in essence is human in their own view, perceiving themselves as people. People, also non-human ones, have agency and intentionality and are thus subjects and not mere objects. In effect, there is thus a close relatedness between Amazonian Indians and their environment, a unity of life, including spirits and what Westerners perceive as abiotic parts of nature. With respect to the Makuna people of northwestern Amazonia, Århem (1996) speaks of a "pact with nature". Within this overarching living universe, food relations play a crucial role, with humans being both potential prey to as well as predators of other species. Human predation of animals and plants is not just an act of necessity but seen "as a generative act through which death is harnessed for the renewal of life" (Århem 1996, p. 200). All beings, all kinds of people, including spirits, play roles in maintaining the whole, and every species has rights and responsibilities in this "cosmic food web". As Århem puts it:

> Human life is geared to a single, fundamental and socially valued goal: to maintain and reproduce the interconnected totality of beings which constitute the living world; "to maintain the world", as the Makuna say. In fact, this cosmonomic responsibility towards the whole – and the accompanying shamanic knowledge – is, according to the Makuna, the hallmark of humanity.
> *(p. 201)*

Thus, there is no idea of humans being superior to other beings. The "pact with nature" is also a "pact of reciprocity" (ibid., p. 191), with humans and animals considered as ontological equals. Those things which are eaten must be blessed before eating, with rites specific to the respective kind of food, in some cases with the support of a shaman (see Århem 1996 for details). An important reason for these rituals appears to be that

> [b]y blessing their food, human beings turn animal-persons into human food and thereby assert their humanity. This shamanic capacity allows humans to overcome the dangers inherent in "nature" while at the same time incorporating the life force it contains.
>
> *(ibid., p. 194)*

Reciprocity means following food rituals but also to provide gifts (e.g. coca) to the "Spirit Owners" of animals. Otherwise, punishment, e.g. in the form of disease, might follow.

Environmental knowledge among Amazonian Indians is high and closely interrelated with religious knowledge (Balée 2003). In his treatment of the Tukano Indians (of which the Makuna are part), Callicott (1994) deals specifically with the ethics and practice of the people. While not diving very deep into the cosmology of the people, Callicott especially emphasises how particular practices of this people – from birth control to hunting regulations, based on their cosmology and traditional knowledge – appear as similar to modern ecological insights, such as (eco)system theory.

For Amazonian Indians the rainforest or other landscapes were never a "wilderness", "pristine" nature without humans, but more a cultural landscape, both from their cosmogony as from their world view and their practices. The different Amazonian cultures are partly hunters and gatherers, partly perform some agriculture, often a mixture of both. It is well-known meanwhile that they actively managed their environment since thousands of years, as they do today (Balée 2003, Mann 2005).

In the context of this book, the interesting thing about Campa, Makuna, and Tukano world views (and similar ones of other peoples) is that it is completely different from and even undermining the received Western ideas of cosmology and of the superiority of humans over nature. Instead, a great network and unity of all that exists is perceived. The kinship idea that is often also put forward by Western conservationists, mostly based on a common evolutionary heritage, here finds an extremely strong expression, by seeing also non-humans as *persons*, and thus human, with "being human" considered as a *condition* and not a matter of species identity. All these persons have agency and intentionality. The result is an attitude of respect and even reciprocity, with humans being responsible for "maintaining the world".

Ideas of respect and reciprocity towards human and non-human nature can be found in many cultures around the globe. To illustrate more in detail how respect is expressed, I will now turn to a second example, namely that of people living in the boreal North of our planet.

Showing respect: native people of northern North America and northern Europe

Native people of North America ("Indians") have often been kind of the prototype of "noble savage" for the 20th-century conservation movement. Their human–nature relationship is indeed interesting for conservationists even though it is sometimes also at odds with some common conservation ideas.

It should be obvious that there is not one "Indian culture" in North America. Even today around 150 different indigenous languages are still spoken in North America (remaining from an estimated 300 languages before European intrusion), mirroring an enormous cultural diversity. I will focus here mainly on the peoples in the boreal zone of the North American continent, drawing largely on the work of Fikret Berkes with the Cree people of James Bay, Canada; I will take some looks aside also at northern European Sami people. Most of these peoples are traditionally hunters, fishers, and gatherers, and the Sami are reindeer herders, today mostly with a mix of income types and subsistence modes. Also their world view and spirituality have been influenced by other views and ideas, predominantly by Judeo-Christian religion. The traditional ways of thinking and living, however, still play an important role in the daily life of indigenous people (Booth 2003, Berkes 2018, Helander-Renvall 2010).

Like for the Amazonian Indians, also boreal indigenous groups see humans and other living beings as intimately related, consider them as fellow beings with agency and self-consciousness. As Booth (2003) emphasises, native Americans did not worship nature but saw every part of it, especially all life, as sacred. From this, a mythological kinship with other living beings follows, also a meaningful personal or even "person to person" (Booth 2003, p. 333) relationship with them and a sense of mutual (!) respect . Like for the Amazonian indigenous people, humans are part of a larger social network, including non-humans.

How does this respectfulness look like in practice? Berkes has worked and lived with the Cree people of the eastern James Bay Area for an extended period of time. He describes three major points for the Cree world view as important for their relation to nature:

(1) it is the animal, not people, who control the success of the hunt;
(2) hunters and fishers have obligations to show respect to the animals to ensure a productive hunt; and
(3) a continued, proper use is necessary for maintaining production of animals.
(Berkes 2018, p. 110)

The first point is quite unusual from a Western point of view. The animal freely "gives itself" to the hunter, but only if certain conditions are fulfilled, especially respect in several forms (point 2). Finally, quite contrary to what most conservationists may declare, there is not only an overuse of animals possible but also an underuse.

To keep the relations with other beings going, several steps of showing respect are to be followed, before, during, and after hunting; several rituals exist for this (see Berkes 2018, p. 115 ff. for more details). It starts with the demand for an attitude of humility:

> The rule about an attitude of humility is both important and universal. Hunters should not boast with their abilities. Otherwise, they risk catching nothing because they are disrespectful of the game.
>
> *(Berkes 2018, p. 117)*

Other procedures of showing respect include a quick and simple killing, making offerings to the animal, butchering and consuming according to rules signifying respect, and last but not least disposing of the remains of the animal in a proper, respectful manner. As Berkes emphasises, it is important to eat what one hunts, with "sport" hunting (without consuming the prey) considered as a transgression or violation of the rules of respect. Offerings are made with, e.g., parts of meat or skin, or today with tobacco.

Offerings, but also the conviction that disrespectful behaviour will lead to unsuccessful hunt, point to the importance of reciprocity in indigenous world views. You cannot only take but you also have to give back, as the others (e.g. your prey) are subjects, yes persons, as well. Animals have an obligation to feed humans and humans must pay respect to them – and use them continuously. *Underuse* is a theme for the Cree and other indigenous groups: a neglect of animals, plants, or the landscape as a whole is as problematic as overuse – there is obligation of humans to use and thus care for the land. That includes also rotations of hunting areas and resting of the land, accompanied and steered by well-established traditional ecological knowledge about the area.

Even stronger than the Amazonian example, Cree worldview (and that of many other indigenous peoples) does not only involve using nature as parts of a good relationship to the world, but even demands it, a point that distinguishes this and many other indigenous human-nature relationships fundamentally from that of wilderness ideas.

The land and the notion of place, not only individual plant and animal species, play an important role for North American Indians as well as for many other indigenous cultures. "Place" is not just territory but something that is born out of the attachment and history of people with a particular area. Tuan (1977) writes: "What begins as undifferentiated space becomes place when we endow it with value" (cited in Berkes 2018, p. 262). Place confers identity, not only in indigenous societies. Place in this sense, I would say, in fact is not as much a topographic but a relational concept. For the Sami in northern Scandinavia, e.g.,

Helander states that places "gather experiences and engage human and non-human persons in various activities. By doing so, they (places) become themselves social beings and actors" (Helander-Revall 2010, p. 49).

The intimate relationship between the land and spirituality and its extreme importance for indigenous groups' identity has been stressed very much by North American Indians as by other indigenous peoples (see especially the following example from Australia).

However, Booth (2003, p. 335) warns:

> Such an indelible bond with the land, however, is clearly a deliberate construction by the tribes. Indian groups have been migrating across the North American continent for perhaps the last 40,000 years. Some are very recent arrivals; the Navajo only migrated into the American Southwest from Alaska and the Yukon about 600 years ago. They are not "native" to the area. They have, however, reconstructed their religion to reflect the new land in which they dwell, in part by borrowing from earlier residents such as the Hopi.

This does not diminish the importance of the (sense of) place for indigenous people, it instead shows that traditional ecological knowledge and the practices and relations people establish with the land are not static but that learning processes occur.

Australia: country as narrative

Relations between humans and their environments, not the least to specific places, are often encoded in the form of narratives. The world view of Australian Aborigines may be the most expressive use of narrative for describing these relationships.

An estimated 250 indigenous languages exist on the Australian continent, and a diversity of different environments, in which people have been living for almost 50,000 years – traditionally as hunters and gatherers, – leading to a diversity of specific cultures and practices. In spite of this, there are also many commonalities. A crucial one is Australian Aborigines' relation to the land, or as they prefer to say "country". Country is linked to identity and the basis of this is not alone a long habitation in and familiarity with particular locations but has religious roots. These roots tie people to country, namely via stories of the creation of the world. Aboriginal creation myths describe how the world came into being. At first there were only the sky and the earth, the latter being barren and featureless. Out of this earth, during the "dreamtime", creator beings, heroic ancestors emerged. As Breeden (2010, p. 15) describes it for the Anangu people of central Australia:

> Some came as humans, others had animal forms and others again changed from one to the other. These creator beings moved all over central Australia, building up mountains, gouging out rivers and digging waterholes.

The journey of these ancestor beings, or their paths, respectively, are called "dreaming tracks" or "songlines", as the ancestors are said to have sung the country into its current existence during their journey. The spirits of the ancestors returned to the land and became totemic ancestors, the spirits themselves still existing in specific places and objects (e.g. trees, rocks, water holes) along the songline. This cosmology is preserved and re-enacted by stories, songs, dances, and paintings. One could thus say that the aboriginal world view is essentially of a narrative character. The songlines provide meaningful relations between human and humans, between humans and non-humans (including non-living things such as rocks), and between humans and the spiritual world. On behalf of this, many parts of the country are sacred and aboriginal identity is strongly anchored in specific places (Strang 2005). The stories also provide codes of proper behaviour. By means of their rituals and practices, encoded in aboriginal law, humans have to perpetuate the old lessons learned from the ancestors to keep the world running as it is, or secure its "rejuvenation", respectively (Callicott 1994, p. 177). As Veronica Strang put it (for Aborigines of the York peninsula):

> The message of the ancestral stories is that the clans are intended to relive the lives of the ancestral beings. Human existence is therefore presented as a continual recycling of spiritual forces out of the land and back into it.
> *(Strang 2005, p. 44)*

Songlines often span huge distances, can branch, and even cross language boundaries (Kohen 2003; see, e.g. the wonderful book about the *Seven Sisters'* songline, edited by Margo Neale 2017).

The structure and the rules of social relations (kinship) in Australian aboriginal societies are extremely complex and their elaboration is far beyond the scope of this book. Some of these relations stand out when it comes to human–nature relationships. Like in many other indigenous communities around the world, social relations are not limited to other humans but also reach beyond. As one important relation, persons are connected to specific totems (plant, animal, or other natural object). Totems are believed to be the descendants of the heroes from the Dreaming stories. Totems exist on several levels of society (nation, clan, family group, and individual).[4] The specific totem defines the roles of a person and its relation to other beings in the world, also its responsibilities (Callicott 1994, Strang 2005). Persons whose totem is, e.g., emu or red kangaroo have responsibilities to be custodians for the fate of these respective animals. They are not allowed to kill and eat any of the animals belonging to their totem, with only rare exceptions, and they have the duty to perform the received "increase ceremonies" to ensure the continued existence and plentifulness of the species (Callicott 1994, Kohen 2003). In effect, hunting of any totemic species will be done only by part of the people, thus lowering hunting pressure. Also, there are additional rules that restrict hunting to specific times and do not allow it at all in sacred places connected to totemic species (Kohen 2003).

Land is not "owned" as private property, but *taken care of*. One might perhaps, if I understand this correctly, say that it is not the land that is owned by individuals or groups but parts of the Dreamtime stories that are owned, going along with responsibility to keep them alive and guard their rules. Land, or country, is not seen as passive but, as in many other indigenous traditions, active and sentient, "a social mirror that acts as an equal partner in the human-environmental dialectic" (Strang 2005, p. 46). A reflexive discourse about "nature" as opposed to humans and their culture does not exist, both are highly integrated. That does not mean that aboriginal people refrain from strongly interacting with and modifying the land. Like most other indigenous people, Australian Aborigines not only used but really *managed* the land, changing it substantially (Flannery 1994). The most conspicuous practice is the common use of fire (Kohen 2003), a tool also used by many other peoples around the world (Pyne 1997).

In spite of their seemingly constant ceremonial relations to the world, to country, constantly continuing the mythic past into the future, aboriginal culture is not static. In its long history, it adapted to major environmental changes, such as sea level rise in what is now Australia (Nunn 2020). Also, it does not lead to a static view of "country" and its components. Trigger (2008) has explored the attitude and relations of Australian aboriginal people to introduced plants and animals. While these attitudes are varied and ambiguous among different people, Trigger demonstrated that there is no simple link between aboriginal nativeness and identity and a rejection of introduced species as "not belonging" to country, as it is sometimes assumed. There is more often an openness to new experiences and phenomena, and sometimes, over time, a kind of acceptance of new species as part of "country". Thus water buffalo (*Bubalus bubalis*), introduced from Asia in the first half of the 19th century, or feral animals like horses and cats, but also some exotic plants, in many places have been actively incorporated meanwhile into aboriginal tradition – without neglecting some negative effects they may have on native ecosystems. Not only are these exotic species used in everyday lives, they are also the subjects of songs and dances; there are even reports of new myths and Dreamings regarding, e.g., buffalo and feral cat. In some cases, introduced species have actually found "cultural space for recognition of what we may understand as an 'emergent autochthony' among these plants and animals" (Trigger 2008, p. 642) and become part of "country". This also causes conflicts with classical Western conservation, namely in those cases where forms of "nativism" (see Box 4.1) prevail. While exotic species are frequently seen by conservationists as something that spoils the naturalness of areas and thus must be combatted, aboriginal society is, to paraphrase David Trigger, more "intellectually generous" towards these organisms and open to change. Traditional ecological knowledge, in world view as in practice, is subject to change, responding itself to environmental and social changes. Learning processes are part of it (Berkes 2018).

Traditional human–nature relationships revitalised? The Satoyama Initiative in Japan

The highly industrialised society of today's Japan maintains many traditions. During the long and complex history of Japan, human–nature relationships were shaped by different religious and philosophical traditions, in particular Shintoism and Buddhism, but also Taoism and Confucianism, with Shintoism being the most "indigenous" one. In consequence, there was not one clear kind of traditional human–nature relationship and associated practices through the centuries. The different traditions (often being mixed) in part led to a reverence for nature perceiving parts of it (especially mountains or sacred trees, as in Shintoism) or the whole of Japan as sacred, and a compassion with all living beings, as in many traditions of Buddhism. Not all Buddhist schools in Japan, however, shared this compassion, some having no objections to killing and eating animals. Some Buddhist schools in fact had "otherworldly tendencies" (Tucker 2003) with little esteem for nature. Morris-Suzuki, in her analysis of the history of human–nature relationships in Japan concluded "that Japanese attitudes to nature, even in pre-industrial society, were both diverse and dynamic" (Morri-Suzuki 1991, p. 97). Part of these attitudes amounted to a veneration of nature and put an emphasis of humans as being indivisible parts of nature, but also other philosophies existed, emphasising the special and prominent role of humans *within* nature, exploring and "opening up" nature to the benefit of humans. In 17th/18th century Japanese philosophy "opening up of things" meant both to "reveal the nature of things" – not the least in search of principles for human morality – but also "developing/making use of the natural world" (Morris-Suzuki 1991, p. 84). These views were influential in the practice of dealing with nature. However, "[c]areless use of these resources – a use not based on the 'laws of heaven and nature' – was wrong because it made people (and the state) worse rather than better off" (ibid. p. 96).

It should be mentioned that the sacredness of nature, when brought to the fore, was often very political, serving inter alia to legitimise the special status of Japan as a country and a geopolitical unit or even particular governments and rulers (Tucker 2003).

In any case, there was always a strong relation between "nature" and culture in Japan and not the clear separation between humans or culture and non-human nature. In describing the *satoyama* concept) (see later), Knight (2010) speaks about an "encultured nature" (see also Chakroun and Droz 2020).

In terms of language, the English word "nature" is commonly translated into Japanese today by the word "*shizen*" (other pronunciation: "*jinen*"), derived from ancient Chinese "*ziran*". First used in the eighth century, *shizen* originally was used more in the meaning of "naturalness", designating something that is "from itself", i.e. the spontaneous state of things (Tucker 2003, Droz et al. 2022). However, even though the terms were important in Japanese literature over many centuries, Tucker (2003, p. 162) states that:

> it never became "the" conceptual category exclusively signifying nature, materialistic or otherwise, in literature, poetry, philosophy, or religion.

Today *shizen* is used both in the original meaning and also in the (Western) dichotomous sense of that what is without human influence (Droz et al. 2022).

The high aesthetic appreciation, even veneration of "nature" in Japanese culture is well-known and reflected in art and poetry long since. While some traditions point towards a high esteem and respect for nature, several authors have pointed out at what Callicott (1994) calls the "paradox" of a disjunction between these attitudes and the actual, often highly exploitative and destructive practice of dealing with non-human nature. Knight (2004) in the same vein speaks about the "Japanese ambivalence towards nature" (see also Chakroun and Droz 2020). An overuse of forests in Japan is already documented in the 17th century, which, perceived as an "environmental crisis", then was met by measures of reforestation (Tucker 2003). Practices leading to the destruction of natural or semi-natural areas, e.g. trough housing development in peri-urban areas, as common today, thus cannot be attributed to Western influences alone. There were, however, not only traditional philosophies conducive to what is today seen as "conservation" but also some long-established practices.

I want to focus here on one old Japanese tradition of "living in harmony with nature", which is meanwhile promoted officially as "the" Japanese approach to nature conservation, namely the concept of *satoyama*. Satoyama (*sato*: village, *yama*: mountain) refers to a cultural landscape around villages in forested mountain areas.[5] Catherine Knight (2010, p. 422) cites a Japanese dictionary definition as:

> the woods close to the village which was a source of such resources as fuelwood and edible wild plants, and with which people traditionally had a high level of interaction.

Very common in these woods were coppice use and forest meadows, the forest being thus less dense, with more light inside than the unmanaged mountain forests further uphill.

There are various other definitions. Sometimes also the agricultural areas (rice paddies, etc.) are included, all together creating a small-scale mosaic of land uses, sometimes even the village is seen as part of *satoyama*. Some modern authors almost equate *satoyama* with any traditional rural landscape in Japan (see Knight 2010, Jiao et al. 2019 for overviews of definitions). Depending on the definition, different estimates are given for the area of Japan that is *satoyama*-land, namely between 18 and 67% (Jiao et al. 2019, Chakroun and Droz 2020). The managed and semi-managed lands normally harbour a rich flora and fauna and are seen as an example of a "nature-harmonious society" (Takeuchi 2010).

The *term satoyama* has been used since 1759 but archaeological evidence suggests that the practice exists since several hundreds, if not thousands of years (Knight 2010). It is thus clearly a traditional idea – or at least practice – of human–nature relationship. The areas have been providing the people of the mountain villages with many resources, from firewood and charcoal to berries and mushrooms (most prominent today the *matsutake* mushroom). As in other areas, where, e.g., fire was applied for management, many species – both those used by people as other species – benefited from intentional disturbances (here especially irrigation and the creation of rice fields) and existed in higher abundances than they would have been in an "untouched" denser forest (see Jiao et al. 2019 for more details). Overall, these areas are known for their high species diversity.

It appears that the lighter forest also served as a kind of buffer zone towards large animals from the adjacent mountain forests (e.g. bear, wild board, and monkeys) that are dangerous or at least detrimental to human enterprises such as agriculture. According to Knight (2010, p. 425):

> Owing to the constant presence of human beings, and the higher level of light and visibility through the vegetation in satoyama woodlands, these areas were not generally frequented by larger forest-dwelling creatures.

This reduced potential human–wildlife conflicts, which have increased since the deterioration of *satoyama* areas.

Especially after WWII, with increasing urbanisation, diminishing rural population, as well as with increasing use of artificial fertilisers and new energy sources that reduced the need for *satoyama* products such as charcoal and firewood, many *satoyama* areas became neglected and degraded, or destroyed. Abandoned sites were subject to natural succession and thus developed into dense forest, many other sites were sacrificed for urban and suburban housing areas. During the 1960s, this became perceived as a problem and many initiatives were developed to maintain or revitalise *satoyama* areas (Takeuchi et al. 2003, Takeuchi 2010). While the practices of *satoyama* are well-documented, I found almost nothing about philosophical and/or religious underpinnings specifically of the people who originally or at least traditionally created and maintained *satoyama*. This is probably due to the fact that *satoyama* and *satoumi* originally were simply descriptive terms for particular types of cultural landscapes and not general ideas or even philosophical concepts.[6] In the current literature, both from scientists promoting the concept as from NGOs and government agencies, there is time and again the assertion that *satoyama* is a model for "nature-harmonious society" (e.g. Takeuchi 2010, or the website of the "*satoyama* initiative": https://satoyama-initiative.org/). But there is not much explanation what that means, if it is only a matter of practice or also a matter of particular attitudes and values. The latter are also difficult to entangle as the concept, in the way it is propagated today, has been variously transformed, extended, and sometimes redefined to serve current purposes of conservation and national politics. But these transformations also make it an interesting example in the context of this book, namely for the question if and how traditional ideas (or at least practices) of nature and human–nature relationships can be used for current conservation approaches. What is obvious is that the relation to nature that is praised here as a harmonious one (Takeuchi 2010) is not one of distance between humans and non-human nature, of excluding humans. *Satoyama* has been created by active human intervention, which must be continued in some way to maintain it (even by mimicking old practices, as done also in European cultural landscapes). It was clearly an important practice for sustained subsistence of the villagers that maintained *satoyama* areas, which does not mean that was all to it. It may be supposed that also aesthetic and other non-instrumental relational values were involved. Today, with the material outputs of *satoyama* largely substituted by modern society through other means, the latter values become the most prominent ones. Knight describes, based on results from an empirical study, that it is more the symbolic, nostalgic appeal of *satoyama* that dominates the public view of it, even much more than, e.g., wildlife protection in adjacent "natural areas". Here it is often linked to the concept of *furusato* (literally ancient village), which promotes the idea of belonging to a traditional village or landscape (Bulian 2021), coming close to the German concept of *Heimat* (see Section 2.2) – with all its ambiguities. In line with this,

> Kitamura (1995, p. 116) proposes that it is the nature of the *satoyama*, rather than the "wild nature" of the *yama* (mountainous forests) with which the Japanese have formed an affinity over the centuries. Further, he suggests that it is the Japanese interaction with this category of half-cultivated nature that has been instrumental in shaping the Japanese view of nature (*shizenkan*) and that has led to the much-vaunted Japanese "love of nature".
>
> *(Knight 2010, p. 436)*

As mentioned, the *satoyama* idea has, however, been heavily extended, transformed, and even instrumentalised. In the context of bringing it into international conservation scene, it has been used and conceptualised in a more scientific parlance as a "traditional social-ecological production landscape" (https://satoyama-initiative.org/, see also Chakroun and Droz 2020). Beyond that it was extended in scope and, since around 2010, used for national showcasing by promoting it as the "Japanese way" of nature conservation that may be a model not only for Japan but worldwide (Knight 2010). However that happens at the price of subsuming many landscape systems labelled as "*satoyama*-like landscapes" under an extended (or one may say diluted) *satoyama* concept, more and more detached from its specific traditional context. It thereby may lose much of its potential power and may not do justice to the many specific other cultural landscapes, which in their specific local or regional contexts, can also serve as examples for traditional human–nature relationships and the practices supporting it (see, e.g. von Droste et al. 1995).

BOX 4.2 JUDEO-CHRISTIAN RELIGION AND NATURE

Judeo-Christian ideas on nature are not found as explicit arguments in the mainstream of conservation (but see Kirchhoff 2023). Historically, however, they played an important role in shaping many understandings of nature and human–nature relationships and at least implicitly they still do, being the common cultural background of most Western societies. The cosmology narrated in the book *Genesis* in the Bible is rather well known to most people. Theologians today do not consider the creation in six days as something to be understood analogously to scientific explanations. Especially the first of the two (!) creation stories, written around the time of the Babylonia exile, served in the first place as a narrative to confirm that God is reliable, does what he promises, is almighty, and leads his people through history. What is more interesting in the current context is the way the relation between humans, nature, and God is characterised, and with it the role that is attributed to humans within God's creation. In the first creation report, God creates humans "in his own image" (Genesis 1, 26)[7] and says to them: "Be fruitful and multiply and fill the earth and subdue it, and have dominion over the fish of the sea and over the birds of the heavens and over every living thing that moves on the earth" (Genesis 1, 28). In the second creation narrative (Genesis 2, 4–24) then, God created the world, then "planted a garden in Eden" (Genesis 2, 8) and put humans into it, "to work it and keep it" (Genesis 2, 15) – a somewhat different mandate for humans than to "have dominion" over non-human nature. Especially the mandate of "dominion" has been interpreted in very different ways – hinging in part on the precise understanding of the original Hebrew verbs of ruling in the book *Genesis* in their historical contexts. Famously, historian Lynn White Jr. in 1967 saw in this text, together with the ideas of the Enlightenment, the "historical roots of our ecological crisis". He made the Judeo-Christian world view responsible for current environmental problems, as creating a dualism between humans and nature also because Christianity "insisted that it is God's will that man exploit nature for his proper ends" (White 1967, p. 1205). This thesis has been discussed fervently and repudiated by many authors, up to this day (see LeVasseur and Peterson 2017). Notwithstanding some exceptions, such as St. Francis of Assisi, as well as a "re-sacralisation of nature" by physico-theologists in the 17th century (Kirchhoff 2023), there was in fact no strong affinity between Christian beliefs and an attitude of caring for nature through most of its history (see Groh 2003, Jenkins 2008 for this history). It is, however, certainly also too easy to see Christian cosmology as the main reason of the ecological crisis. Since the onset of the conservation and environmental movements, the role that a Christian world view can or should play for conservation has been the subject of much debate and is still further developed (e.g. Northcott 1996, Jenkins 2008, Conradie et al. 2014). It should be noted that in Christian reasoning,

> duties towards nature are indirect only. Christian stewardship of nature is not a stewardship with responsibility towards nature itself but towards God, with the mandate to keep his creation in good order.

With the example of Japan and *satoyama* we have come almost full circle, back to the Western ideas of nature and the overall question of how to link them with traditional and even indigenous ones.

What I have done in this section is to describe indigenous and traditional ideas of "nature" and human–nature relationships. With the exception of the current use of *satoyama*, however, these are not *conservation* concepts, but more general ways of living in the world, which, from a Western conservationist perspective may provide some potentially interesting ideas. Alcorn (1993, p. 425) even states:

> To my knowledge, there is no direct translation for the word "conservation" in any non-European language. It is generally translated as "respecting Nature," "taking care of things," or "doing things right."

Under the broad definition by Sandbrook (2015), with which I started the book, however, these latter expressions would fit quite well as "conservation".

Imposing Western ideas of conservation unquestioned unto people with other cultures is highly problematic (see Section 5.2), but it is also problematic to use concepts from other cultures unreflectively as "conceptual resources" (Larson 1987) for improving our own environmental ethics and human–nature relationships. Gerald James Larson warns against what he calls the "fallacy of disembedded ideas" (Larson 1987, p. 156). This means that ideas and concepts are not independent of the cultural context in which they are used, of their framings, especially the functions they have in this context. Simple cherry-picking of ideas and trying to fit them in our own concepts should thus be treated with healthy scepticism.

A consideration of the ideas and practices of human–nature relationships in other cultures is nevertheless of high value. It can help to reflect on what is common in some ideas but neglected in the mainstream of conservation or even society as a whole, to see which traditions exist and which perhaps have been lost and may be revitalised. Again, one has to be careful to not see commonalities or exotic differences too easily, taking things out of context, which will be one issue in the case study in Section 4.4.

First of all, it appears that two of the narratives that are popular in conservation are totally alien to indigenous human–nature relationships. Neither "Nature apart from humans", as represented by a strict wilderness idea, nor a dominance of "Useful and necessary nature", as represented by an extreme version of the ecosystem services concept, are to be found in indigenous traditions. Of course people need and use nature, but in a completely different context of relation to non-human nature, not perceiving of nature as a *mere* resource.

Many people might not go back to religious or spiritual roots for conservation, but the notion of *respect* for others, both human and non-human, is not in any way alien to Western culture, both in everyday life as in philosophy. In terms of respect for nature, one could mention, e.g., Albert Schweitzer (1923/1990) with his philosophy of "reverence for life" ("*Ehrfurcht vor dem Leben*") or environmental philosopher Paul Taylor (1986/2011) and his "respect for nature". So here we would have a candidate for connecting to non-Western human–nature relationships.

Related to respect is a perception of *human embeddedness in nature* which permeates many traditional world views. Accepting this does not necessitate to give up a distinction between humans and nature, but it means to understand this distinction in a more relational and less hierarchical way, not putting human interests always above that of the rest of nature. Many existing conservation concepts and narratives are in principle embracing this.

The ideas of human embeddedness in nature, together with respect towards it, can lead to an attitude and practice of *caring for nature/country/the land*. There are multiple ideas within Western society, and even conservation, on which to draw (see Jax et al. 2018). On the conservation side, Aldo Leopold's Land Ethic has been interpreted in this sense. Also "deep ecology", according to its founder, Arne Naess, should aim at "extending care to humans and deepening care for non-humans" (cited in De Jonge and Whiteman 2014, p. 449). Another recent and important "tradition" that should be mentioned here is eco-feminism (e.g. Warren 2015).

The notion of *personhood of nature*, or parts of nature, as embraced by some indigenous cultures, is more difficult to connect to with Western ideas, especially on a philosophical, moral level, at least beyond sentient animals. It has, however, been taken up as a mainly legal concept more recently, attributing in some cases legal rights to, e.g., rivers – an interesting but difficult discussion into which I cannot delve further (but see Gordon 2018, Page and Pelizzon 2022).

To sum up, dealing with non-Western or non-mainstream ideas of nature and human–nature relationships is important in two ways. First of all it is a matter of respect and justice towards other people, their world views and cultures. We must not simply take our own ideas of nature conservation for granted and as based on universal ideas. Even we consider, as most of us do, conservation as a matter of high urgency, we should not simply impose our own ideas – and their major practical implications – on other people, who might have completely different understandings of what makes a good human–nature relationship. I will discuss this in more detail in Chapter 5. Second, we can learn by opening up the discussion about our own self-understanding asking what it means to be human. The diversity of existing ideas of human–nature relationships and practices related to it may inspire new ideas and "excavate" old ones.

While integration of different knowledge forms as well as an intensified discourse between different world views is desirable, one should, however, resist to searching for an overall integration of these other world views into one general idea of conservation. Sometimes both respect for other cultures (and different ideas in our own culture) and our own interest may be served better by keeping these different concepts in a productive tension (see Section 4.4).

For non-scientific approaches in general (e.g. religious ones) and those from other cultures specifically, narratives can be a useful tool to find some common ground that goes beyond a naive "translation" of indigenous knowledge, concepts, and world views into scientific ones. I will discuss that in Section 4.5.

The following section on the IPBES conceptual framework will demonstrate how difficult it is to integrate different world views for conservation. It will also show that and how concepts of nature can have political relevance.

4.4 Case study: How concepts of nature become political: the discourse on the IPBES conceptual framework

The Intergovernmental Platform on Biodiversity and Ecosystem Services (IPBES) was established in 2012, by almost 140 states. IPBES is sometimes also dubbed the "World Biodiversity Council" and is, roughly, for biodiversity what the IPCC is for climate, however, with a stronger emphasis on stakeholder involvement and the inclusion of various knowledge forms (Beck et al. 2014, Löfmarck and Lidskog 2017). As the name indicates, it is an independent, intergovernmental institution with the overall goal to "strengthen the science-policy interface for biodiversity and ecosystem services for the conservation and sustainable use of biodiversity, long-term human well-being and sustainable development" (https://ipbes.net/about). The current work programme (until 2030) has six objectives. The most well known of these is the assessment of knowledge. Several assessment reports have already been published (https://ipbes.net/assessing-knowledge). Some of these assessments are on the status of biodiversity, partly global, such as the Global Assessment Report on Biodiversity and Ecosystem Services, some on more regional levels. Yet other assessments have more specific themes, such as the Values Assessment or the Land Degradation and Restoration Assessment. Another objective of the work programme is "strengthening the knowledge foundations". There is a special emphasis on knowledge forms different from the dominant Western, scientific knowledge, not only including social and engineering sciences but very importantly also "recognizing and working with indigenous and local knowledge" (https://ipbes.net/o3-strengthening-knowledge-foundations; see also https://ipbes.net/indigenous-local-knowledge and UNESCO 2013). This is an important step forward beyond the mainstream of conservation and conservation biology, which will be challenging to the conservation community. Further objectives of IPBES are, inter alia, capacity building for science-policy interfaces on biodiversity and ecosystem services, and to support policy. Given the complexities of the tasks and the insight that also societal aspects, institutions, and governance had to be incorporated to serve its objectives, IPBES early on developed a *Conceptual Framework* (CF)

Western and non-Western ideas of nature and nature conservation **195**

Analytical conceptual framework

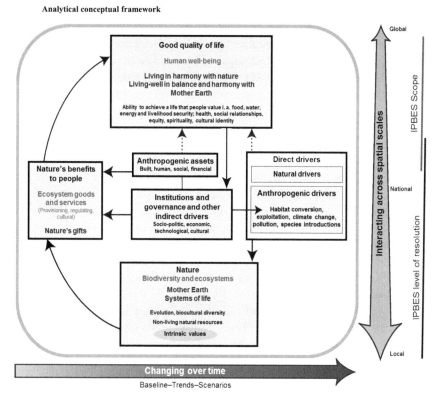

FIGURE 4.1 The IPBES Conceptual Framework.

Source: UNEP 2014 (Decision IPBES-2/4); see text. Reprinted with permission of the IPBES secretariat.

to provide coherence between its different tasks, to display the relations between different concepts and actors, between "the natural world and human societies" (UNEP 2014), and to provide a simplified model thereof (Figure 4.1).

I will not go into every detail of this framework (see UNEP 2014, Díaz et al. 2015a, 2015b for detailed elaborations) but, after a short general explanation, will focus on one aspect which is of special interest for this chapter, namely how different knowledge forms and, with it, different ideas of human–nature relationships are handled. The framework's approach to this issue is rather unusual and is very illustrative for discussing potentials and problems of linking Western perspectives on nature and nature conservation with non-Western approaches.

The IPBES conceptual framework

The scheme is composed of six boxes, depicting the main elements of nature and society in the focus of IPBES, and arrows that connect them in terms of interactions.

The box forming the centre of the scheme is – quite unconventional – not "nature" but "institutions and governance and other indirect drivers", thus emphasising the crucial role of society and its institutions for anthropogenic changes in biodiversity and ecosystems – both negative and positive. Three of the boxes, namely "nature", "nature's benefits to people" (meanwhile renamed as "nature's contributions to people"; see Section 3.4), and "good quality of life" are labelled with different terms in different colours (in the original; in Figure 4.1 displayed by different shades of grey). This "colour code" is meant to display for each box the Western, science-based concepts (in green, light grey here), related concepts from other (e.g. indigenous) knowledge systems (blue, dark grey here) and concepts considered to be "inclusive of all these world views" (UNEP 2014) in black (bold black here). Thus "Nature" as the overarching term is represented by "Biodiversity and ecosystems" in the Western world view, while it is indicated by "Mother Earth" "Systems of Life" for other, especially indigenous knowledge systems and world views.

Díaz and colleagues (2015b, p. 4) referred to the framework and its colour coding as a "'Rosetta Stone' for biodiversity concepts that highlights the commonalities between very diverse value sets and seeks to facilitate crossdisciplinary and crosscultural understanding". First of all, displaying the main elements of the different world views about interactions between human societies and nature by three different colours is a real progress – instead of simply resolving the tension in one or the other direction, mostly the scientific one, as it was planned in the first drafts of the framework (Borie and Hulme 2015). It is also eventually bringing the debate about indigenous and local knowledge (ILK) into the mainstream of conservation biology and thereby – in principle – challenging many established conceptual approaches. But is the IBPES Conceptual Framework the long-sought stone of different languages that eventually allows to "translate" between the Western scientific approach to nature and non-Western approaches? Or is it – as a British colleague of mine responded when hearing about the metaphor – instead a Procrustes Bed, where the "non-Western" approaches are forced into an uneasy structure? There is no doubt about the good will and energy of those who have developed the scheme, and I am aware that it is very much a product of political negotiations, but one can doubt for good reasons that it can really fulfil the purpose of integrating the different knowledge systems.

Beyond other critiques that have been uttered in relation to the CF (e.g. Maier and Feest 2016), there are two major problems in the context of this chapter: first of all, the colour coding lumps together both all "Western" knowledge and philosophies but, even more important, also all kinds of local and indigenous knowledge (Borie and Hulme 2015, Löfmark and Lidskog 2017). Díaz and colleagues provide at least a short list with some of indigenous world views, next to the Andean "Mother Earth and Systems of living", which found their way into the framework as kind of placeholder:

> the concepts of sẽ nluo´–wa` nxia` ng (vast forest and every manifestation of nature) and tien-ti (Heaven and Earth) of Taoism shared by East Asian peoples,

and concepts of the land encompassing non-human living organisms, living people, ancestors, deities and their shared histories in the South Pacific Islands (e.g. fonua, vanua, whenua, ples).

(Díaz et al. 2015a, p. 4)

But even if we take only one expression of indigenous perspectives, namely that of Mother Earth, promoted by Bolivia, which is used in the CF, the question is as to whether the words in different colours do really mean "the same" – or at least similar ideas – that can be subsumed as "inclusive" under the terms in black, or if they remain (in stark contrast to the image of the Rosetta Stone), strongly incommensurable. Do "Mother Earth and Systems of life" on the one side and "biodiversity and ecosystems" on the other, or at least "Mother Earth and systems of life" and "nature", refer to the same thing? I think they do not and they especially refer to different narratives of human–nature relationships.

The development of the IPBES conceptual framework

To approach this, it is interesting to take a look into the development of the IPBES–CF and also to scrutinise at least those Andean notions of "Mother Earth" and "living well" that gave rise to the wording of the blue coding in the CF.

Starting with the formal IPBES document itself (Decision IPBES-/4, UNEP 2014), the description given there provides little help as they do not give any detail what "Mother nature and systems of life" mean and imply:

> "Nature" in the context of the Platform [IPBES] refers to the natural world with an emphasis on biodiversity. Within the context of science, it includes categories such as biodiversity, ecosystems, ecosystem functioning, evolution, the biosphere, humankind's shared evolutionary heritage, and biocultural diversity. Within the context of other knowledge systems, it includes categories such as Mother Earth and systems of life. […] Nature contributes to societies through the provision of benefits to people (instrumental and relational values, see below) and has its own intrinsic values, that is, the value inherent to nature, independent of human experience and evaluation and thus beyond the scope of anthropocentric valuation approaches.

While the Western notion of "nature" as biodiversity and ecosystem is elaborated somewhat, with almost exclusively scientific concepts, the concepts of other knowledge forms ("Mother Earth and systems of life") are just mentioned but not detailed further. In a journal paper by Díaz et al. (2015a), describing the then new framework, at least some very brief qualifications of these concepts (e.g. "Mother Earth") were given.

Borie and Hulme (2015) have documented the development of the CF within IPBES in much detail. The first drafts of the framework only referred to what is now the green part of the CF, the Western scientific concepts (see figure 3 in

Borie and Hulme 2015). This kind of depicting the tasks and the main elements of IPBES was heavily criticised by the Bolivian delegation at the IPBES meeting in 2013 in Bonn, with support from other South American but also some Western states. In a written statement (Plurinational State of Bolivia 2013)[8] it was emphasised that this terminology did not reflect the world view of many "non-Western modern societies, indigenous people and local communities" and their relationships with nature. The framework should therefore be "enriched" to account for these.

The Bolivian document sees the Western world view and the first framework draft as one "completely biased towards a particular vision of biodiversity which is the one related to the commodification of nature" (Plurinational State of Bolivia 2013, p. 2). Beyond that, and connected to it, it even sees the idea of "nature" as not appropriate to the (South American) indigenous approach:

> In this context, the concept of Mother Earth is completely different than nature. Mother Earth is a living system or living being. This would imply saying that is nature considered as a living being with specific "rights", paralleling the "human rights". In conclusion, Mother Earth is "our mother" and therefore is not an object to be exploited by human beings.
>
> *(ibid., p. 7)*

The context to which the above quote refers is the one of "living well" (in balance and harmony with Mother Earth) and it is explained in the Bolivian statement as:

> Living-Well means living in balance and harmony with everybody and with everything, where the most important thing is not the human beings but the life. Living-Well is living in community, in brotherhood, in complementarity; it means a self-sustained, communitarian and harmonic life.
>
> *(ibid.)*

The concepts of "living well" (*buen vivir*) and Mother Earth (*Pachamama*) are meanwhile parts of the constitutions of both Bolivia as well as Ecuador (since 2009 and 2008, respectively).

These non-Western concepts were seen by some Western scientists and conservationists not as something completely different from their own ideas, the more as the concepts of biodiversity, ecosystem services, and human well-being are often interpreted much broader than serving merely utilitarian purposes, even less embracing a commodification of nature (see Section 3.4). According to Borie and Hulme (2015, p. 493):

> For the ecosystem services community, ecosystem services appear predominantly as an epistemic notion, but which is also a pragmatic way to frame biodiversity-issues. There is nothing intrinsically suspect about ecosystem services. It is a

concept for linking scientific knowledge on biodiversity with policy-making processes.

The "pragmatic way to frame biodiversity-issues" hints to the strategic use of the ecosystem services concept, as an attempt to better address political and business audiences.

The whole Western terminology is also linked to that of the Millennium Ecosystem Assessment (MA), thus providing consistency and continuity. Even if it may not fully be justified, given the actual breadth of values promoted by the MA and by the ecosystem services concept (see Section 3.4), for those who contested this terminology it was far from neutral and merely "pragmatic", but clearly connected to (an extreme version of) the narrative that I described in Section 3.3 as "Safeguarding necessary and useful nature", bringing about a whole system of utilitarian thinking and the dominant economic approach in Western society – a completely different perspective on human–nature relationships than the one for which the expressions "Mother Earth" and "living well" stand.

The suggested alternative terms and concepts were eventually taken up in the final version of the CF as adopted by the IPBES plenary. It should be obvious, already from the statements of the Bolivian document, that different colour codings are not simple "translations" of the same or at least similar ideas that can be integrated by using "inclusive" terms in black. There are fundamental differences, and a tension between the "Western" notions and that of the "other knowledge systems" will remain. Such a tension, in principle, is not a bad thing. A look at the background of the main concepts promoted by the Bolivian delegation, however, shows how strong the differences really are, and, in addition, how complex and diverse ideas on living well and Mother Earth are even in South America.

Buen vivir *and Western human–nature relationships*

The Bolivian document states that the idea of "living well" is "originated in the vision of indigenous peoples worldwide" (Plurinational State of Bolivia 2013, p. 7). There are several movements in South America promoting *buen vivir* (or *vivir bien*) and the relation to *Pachamama*. Gudynas (2011) speaks of a "plural concept" and Escobar (2016) even characterises the various South American epistemologies as "pluriverses", a diversity of worlds, which he contrasts with the "One World World" as the dominant Western world view that attempts to shape everything in its way. Building often on indigenous traditions, the movements share a critique on the Western ideas of development and linear progress and emphasise the value of community (including non-human beings; see also the examples in Section 4.3 for this and the following) instead of an individualistic Western approach to a good life. Also, it rejects both a merely utilitarian view on nature and a dualism between nature and humans. "Nature becomes a subject" as Gudynas (2011, p. 445) emphasises. Thus the "gifts of nature", in the IBPES framework, are something completely different than their counterpart "ecosystem services". A gift – in contrast to a service – emphasises *reciprocity and respect* beyond classical market logic, which is a fundamental feature of human relations to *Pachamama* as understood, e.g., in the tradition of the Aymara in Bolivia (Mamani-Bernabé 2015).[9] Reciprocity can be found in some approaches of *caring for nature* (see Jax et al. 2018*)*, especially those close to feminist ethics, but it is usually not part of any conception of ecosystem services (you take from nature but you normally do not give back; but see Ojeda et al. 2022). It might be noted in this respect that in the CF, the arrow between the boxes nature/Mother Earth and nature's benefits to people/nature's gift, is depicted only unidirectional, pointing from nature to nature's benefits to people and further to good quality of life but not the other way around; reciprocity is not accounted for.

In spite of its old indigenous roots, *buen vivir*, as it is promoted today, is a rather recent and highly political concept with many different expressions (Gudynas 2011, Cuestas-Caza 2018). For example, the Bolivian statement to the first IBPES–CF draft links indigenous ideas to approaches of noble laureate Eleanor Ostrom on the use of commons and ways of governing them to avoid their misuse. Likewise, the ways *buen vivir* and *Pachamama* have been used in the new constitutions of Ecuador and Bolivia are very different (Gudynas 2011). In Bolivia *vivir bien* is related explicitly to various indigenous traditions and concepts (e.g. the Aymara concept of *suma qamaña*) and is largely used as one of several moral guiding principles. Instead, the Ecuadorian constitution, referring to the indigenous concept of *sumak kawasy* as *buen vivir*, establishes the *rights* of nature or *Pachamama*. As Knauß (2018) has pointed out, however, *Pachamama* is equated with "nature" in the constitution of Ecuador. So the use of *buen vivir* and related terms varies a lot, depending on the specific traditions and modern uses. Cuestas-Caza (2018) has analysed very clearly that, e.g., the Quechua concept *sumak kawasy* is not automatically the same as *buen vivir*.

What I wanted to show with this short digression into the roots of the "blue" ("indigenous") terms in the IPBES conceptual framework are two things: first of all, the "green" ("Western") and the "blue" (indigenous/other knowledge system") concepts can certainly not be viewed as mutual translations and equivalent ideas, to be unified under the "black" concepts. The world view behind these concepts and the narratives of human–nature relationships that could be associated to them are often far away from each other, especially rejecting a human–nature dualism and a utilitarian approach towards nature as present in Western economy and also most ideas of "development", still rooted in a largely unquestioned capitalistic logic (see Section 5.2). I am saying "often" (and this is my second point) because the terms designating the boxes in the IPBES–CF, especially for the non-Western knowledge, in a way are placeholders for a large variety of knowledge systems and world views (see examples in Section 4.3), some of which may indeed come close to the "Western" one, many or most of them, however, not. While it is good to display such different perspectives, some authors (e.g. Borie and Hulme 2015), however, see also the danger that the sharp division between Western and other knowledge systems might reinforce classical separations, e.g., between a "rational", enlightened, and "unbiased" Western science on the one side and value-led and partly irrational knowledge systems on the other side. Gudynas (2011, p. 446), however, emphasises that *buen vivir* is not a return to mystical indigenous past but that it presents "precise proposals and strategies", including "reforms in legal forms, introduction of environmental accounting, tax reforms, dematerialization of economies".

Lumping all non-Western approaches into one category to nature is problematic, but the same holds for Western approaches. Conceiving of Western science and more broadly the Western world view as one monolithic block is also painting a problematic picture of what Western science is. Also Western science needs self-reflection, needs critical evaluation of its own premises and (partly unavoidable) values, as e.g. the discussion on the Western concept(s) of nature but also those of the different conservation concepts have shown. Even within Western conservation, which is the major subject of this book, there is not one idea about good human relationships with nature, there is not only one narrative nor one conservation concept that fits all conservation needs and which is embraced by all conservationists. This *multitude of narratives* also within Western conservation is a treasure, not a problem, as are the many different world views and approaches to "nature" in other traditions. This multitude of narratives is in danger to be obscured when the "Western approach" is represented mainly by terms from the ecosystem services (and biodiversity) field. They all have to be, as I will argue in Chapter 5, adapted to specific situations, a thing that Escobar, Gudynas, and others also emphasise for approaches of *buen vivir* and related ones.

I want to mention one last point here concerning the compatibility of different concepts within the IPBES–CF. At least the box labelled "Good quality of life" (overarching), comprising "human well-being" (Western) and "living in harmony with nature/living well in balance and harmony with Mother Earth" (other knowledge systems), may be harmonised by taking in a more complex understanding of human well-being. The expression "human well-being" appears to be taken from the MA framework; it is a complex idea, not limited to merely utilitarian or hedonistic well-being concepts but also including, e.g., good social relations (see Figure 3.4). In the MA, the categories of human well-being were not derived top down from a purely Western perspective but were based on the empirical study "Voices of the poor" (Narayan et al. 2000), collecting statements from poor people in 50 countries all over the world. Moreover, conceptualisations of human well-being such as the capability approach, in the form developed by Martha Nussbaum, open up more and more towards including (relational) values. For Nussbaum, e.g., one of the capabilities which make up a fulfilled human life is:

> Other species. Being able to live with concern for and in relation to animals, plants, and the world of nature.
>
> *(Nussbaum 2011, p. 34)*

However, the human well-being part of the ecosystem services approach is still least conceptualised and has not really been developed to its potential.

The treatment of the case study above focused on the specific ideas of human–nature relationships as expressed in the IPBES–CF. It does not mean to characterise the whole treatment of indigenous and local people in IPBES and the question of how ILK can be integrated or considered on an equal level – which is anything but a simple task. But this is another discussion into which I cannot venture here. For the general treatment of ILK in IPBES, see, e.g. UNESCO (2013), Hill et al. (2020), and also Tengö et al. (2017). Tengö and colleagues also emphasise that the idea of "weaving knowledge systems" – which I think is a wonderful metaphor – is more than translation and integration but that their "multiple evidence approach" also:

> recognises the incommensurability of diverse knowledge systems and the often asymmetric power issues arising when connecting different branches of science with locally-based knowledge systems. [...]
> We view the outcome as weaving—collaborations that respects the integrity of each knowledge system.
>
> *(p. 18)*

As said before, tensions between knowledge systems and world views can be highly productive and should not too easily be resolved. If the IPBES–CF is understood in this way, and if it is used as a starting point and not as end point, it is highly useful and can foster further discourse about different ideas of "nature" and human–nature relationships – and how to deal with it. A danger in resolving such tensions too quickly, and a general challenge to the application of non-scientific knowledge forms, is that non-Western knowledge and ideas are simply used as "conceptual resources" (Larson 1987) and that, at the end of the day, Western science still remains the default and dominant approach to conservation.

4.5 Narratives as a tool to "translate" between "Western" and "non-Western" ideas of nature and human–nature relationships

The difficulties of "translating" between different cultural understandings of proper human–nature relationships and approaches to knowing nature have been emphasised above. Does, then, the use of narratives allow for a bridge between different ideas and world views? I think that they can provide some help, can provide some good points of contact and connection. Narratives need interpretation; they are also set in specific cultural contexts, like knowledge, like values, like even perception. But I nevertheless think that they are often better suited to start communication and interaction than abstract concepts. Narratives come close to everyday ways of describing the world and human relationships within it. They are not just about values embraced (the same type of value can occur within different conservation narratives) but very much also about the self-perception of humans in relation to non-human nature.

Non-Western, especially indigenous and religious world views built very often, and then strongly, on narratives to explain the world and to provide rules of conduct. Narratives are also a way of transmitting traditional knowledge. Of special importance here is that narrative analysis allows to conceptualise also non-human entities as "actants" with specific roles and characteristics – with or without attributing full agency in a human sense to them. In fact, the boundaries between "scientific" narratives (as implicit in conservation concepts) and non-scientific, even mythical ones, is sometimes blurred, such as when it comes to (also "Western") religious justifications for conservation (e.g. in some expressions of stewardship concepts), but also in terms of many metaphorical ideas, e.g. of "nature striking back".

As one more big advantage, narratives may also explain and describe the meaning of complex concepts without defining them. Using them, we may circumambulate an idea, a concept, like a statue, view it from different sides in order to grasp it. Narratives thus are an ideal tool to at least find some common ground that goes beyond a naive "translation" of indigenous knowledge, terms, and world views into scientific ones.

One can ask, for example, do our Western conservation narratives match with narratives from other cultures, and if so, which of them more than others? Is something missing in our narratives?

In terms of practice, indigenous approaches to dealing with the non-human world have similarities with some of the narratives described in Chapter 3. Especially ideas of "Being at home in nature", some form of "Caring for other living beings (and natural features) is part of a good human life" and even "Steering nature in a responsible way", may find resonance in indigenous world views and practices. Some authors (e.g. Berkes and Davidson-Hunt 2006) consider the practices of some indigenous people as very similar to ecosystem management.

Most indigenous narratives of human–nature relationships that I am aware of differ from the scheme for conservation narratives which I developed in Chapter 3 in one interesting point. While a "crisis experience" is constitutive of all Western conservation concepts, such crises – specifically of the relation between humans and nature! – hardly seem to be part of indigenous narratives. This may be partly an artifact of my approach. As emphasised before, my selection of Western ideas of human–nature relationship is limited as I am dealing in this book in the first place with relationships as expressed by conservationists and conservation biologists, not that of the whole of Western societies, which may for some people still go without a sense of crisis. Nevertheless, there are of course instances of crises within traditional cosmogenies. The great deluge, present in many ancient narratives, is a crisis experience, but it is normally seen as moral crisis of human behaviour towards God and towards each other, for which the deluge was a punishment. Berkes (2018) also speaks about "crisis learning" of traditional people, triggered, e.g., by unexpected resource depletion events (see, e.g. Berkes 2018, p. 131ff). He considers crisis learning, however, as the exception, against the dominant background of "environmental understanding learning".

An important connection point, which we might take from Western and non-Western narratives of human–nature relationships is the role of humans within the world (and with respect to what we call "nature") and with it again the understanding of what it means to be human (see, e.g. Peterson 2001). This can broaden our ideas beyond what we take for granted and help to respect other world views and human–nature relationships. But we have to delve into the details in each specific case. Simple generalisations of both Western and non-Western understandings of nature will not help and even less the effort to uncritically assimilate ideas from other cultures into our conservation efforts.

To illustrate the last point and the challenges associated with it, let me close this chapter with a longer quote from a paper of Rachelle Gould and colleagues, who investigated indigenous world views and relations to nature on Hawai'i:

> The Hawaiian language does not have words for values, virtue, morals, or ethics. To address this, we draw heavily on the work of Malcolm Nāea Chun (2011), who also deeply considers this translation issue in his work. Chun has worked for many years to collect and create a written record of Hawaiian principles, philosophies, and practices, both currently and in the past. He draws on a mix of personal experience, other scholarly work, and Hawaiian moʻolelo – stories and legends. He uses multiple moʻolelo – in some cases quite lengthy ones – to develop and describe his points about Hawaiian principles. This mode of communication aligns with Hawaiian ways of communicating about values – the stories demonstrate, but often do not explicitly discuss, values (the centrality of stories as demonstrations of value is far from unique to Hawaiʻi). [...]
>
> Chun implies that in Hawaiian worldviews, values are demonstrated and embodied in practices, rather than discussed. Yet he notes that there are plenty of words for "actions and expressions" that manifest what Western scholarship might call values.
>
> *(Gould et al. 2019, p. 1219)*

Notes

1. "Sie passen ihre Färbung ihrer jeweiligen Umgebung an. Jedes Mal, wenn von 'natürlich' die Rede ist, geht es darum, einen Kontrast ins Blickfeld zu rücken und zwischen dem Natürlichen und seinem jeweiligen Gegenteil zu unterscheiden".
2. Considerable parts of this section are a based on chapter 5 of my book *Ecosystem functioning* (Jax 2010).
3. Cosmogonies are stories or theories about the origin of the world.
4. See www.youtube.com/watch?v=vpHG9V2qJiE for an explanation by an indigenous researcher. Accessed April 20, 2023.
5. A complement to *satoyama* in coastal areas is *satoumi* (*umi*: sea), a village in a cultural seascape. I will focus here on *satoyama* only.
6. Thanks to Laÿna Droz for pointing to that.
7. All quotes taken from the English Standard Version of the Bible.

8   The whole document is available to me and was publicly accessible at: www.ipbes. net/images/documents/Bolivia_comments%20on%20background%20document%20 on%20IPBES%20Conceptual%20Framework.pdf. This source was also cited by Borie and Hulme 2015. However, the link does not work any more, and (as of April 2023) I could find no other source to get access to the Bolivian statement. But see www. genevaenvironmentnetwork.org/wp-content/uploads/2020/05/living-well_pdf.pdf for a brochure by the State of Bolivia on this issue. Accessed April 20, 2023.
9   The acceptance or exchange of gifts has many implications for both donor and recipient. For a description of the concept of gift, here from the perspective of a North American Native and ecologist, see Robin Wall Kimmerer's wonderful essay "The serviceberry" (Kimmerer 2022).

References

Agrawal, A. (1995). Dismantling the divide between indigenous and scientific knowledge. *Development and Change*, 26, 413–439.

Alcorn, J.B. (1993). Indigenous peoples and conservation. *Conservation Biology*, 7, 424–426.

Anderson, J.E. (1991). A conceptual framework for evaluating and quantifying naturalness. *Conservation Biology*, 5, 347–352.

Angermeier, P.L. (1994). Does biodiversity include artificial diversity? *Conservation Biology*, 8, 600–602.

Angermeier, P.L. (2000). The natural imperative for biological conservation. *Conservation Biology*, 14, 373–381.

Århem, K. (1996). The cosmic food web. Human nature relatedness in the Northwest Amazon. In: *Nature and society: anthropological perspectives* (eds. Descola, P. and Pálsson, G.). Routledge, London, New York, pp. 185–204.

Balée, W. (2003). Native views of the environment in Amazonia. In: *Nature across cultures: views of nature and the environment in non-Western cultures* (ed. Selin, H.). Springer, Dordrecht, pp. 277–288.

Beck, S., Borie, M., Chilvers, J. et al. (2014). Towards a reflexive turn in the governance of global environmental expertise. The cases of the IPCC and the IPBES. *Gaia-Ecological Perspectives for Science and Society*, 23, 80–87.

Berghöfer, U., Rode, J., Jax, K. et al. (2022). 'Societal Relationships with Nature': a framework for understanding nature-related conflicts and multiple values. *People and Nature*, 4, 534–548.

Berghöfer, U., Rozzi, R. and Jax, K. (2010). Many eyes on nature – diverse perspectives in the Cape Horn Biosphere Reserve and their relevance for conservation. *Ecology and Society*, 15(1): 18. www.ecologyandsociety.org/vol15/iss1/art18/

Berkes, F. (2018). *Sacred ecology*. 4th ed. Routledge, New York and London.

Berkes, F. and Davidson-Hunt, I.J. (2006). Biodiversity, traditional management systems, and cultural landscapes: examples from the boreal forest of Canada. *International Social Science Journal*, 58, 35–47.

Bird, E.A.R. (1987). The social construction of nature: theoretical approaches to the history of environmental problems. *Environmental Review*, 11, 255–264.

Birnbacher, D. (2006). *Natürlichkeit*. de Gruyter, Berlin.

Booth, A.L. (2003). We are the land: native American views of nature. In: *Nature across cultures: views of nature and the environment in non-Western cultures* (ed. Selin, H.). Springer, Dordrecht, pp. 329–349.

Borie, M. and Hulme, M. (2015). Framing global biodiversity: IPBES between mother earth and ecosystem services. *Environmental Science & Policy*, 54, 487–496.

Breeden, S. (2010). *Uluru. Looking after Uluru – Kata Tjuta – The Anangu way*. Gecko Books, Marleston.
Bulian, G. (2021). The multilocality of *satoyama* landscape, cultural heritage and environmental sustainability in Japan. *Ca' Foscari Japanese Studies. Religion and Thought*, 14, 111–133.
Büscher, B. and Fletcher, R. (2020). *The conservation revolution. Radical ideas for saving nature beyond the Anthropocene*. Verso, London, New York.
Callicott, J.B. (1994). *Earth's insights: a multicultural survey of ecological ethics from the Mediterranean basin to the Australian outback.* University of California Press, Berkeley.
Callicott, J.B. (1996). Do deconstructive ecology and sociobiology undermine Leopold's Land Ethic? *Environmental Ethics*, 18, 353–372.
Chakroun, L. and Droz, L. (2020). Sustainability through landscapes: natural parks, satoyama, and permaculture in Japan. *Ecosystems and People*, 16, 369–383.
Coates, P. (1998). *Nature. Western attitudes since ancient times*. Polity Press, Cambridge.
Cole, D.N. and Yung, L. (eds.) (2010). *Beyond naturalness. Rethinking park and wilderness stewardship in an era of rapid change*. Island Press, Washington D.C.
Conradie, E.M., Bergmann, S., Deane-Drummond, C.E. et al. (eds.) (2014). *Christian faith and the Earth: current paths and emerging horizons in ecotheology*. Bloomsbury Academic, London.
Coscieme, L., da Silva Hyldmo, H., Fernández-Llamazares, Á. et al. (2020). Multiple conceptualizations of nature are key to inclusivity and legitimacy in global environmental governance. *Environmental Science and Policy*, 104, 36–42.
Crist, E. (2004). Against the social construction of nature and wilderness. *Environmental Ethics*, 26, 5–24.
Cronon, W. (ed.) (1995/1996). *Uncommon ground: reinventing nature.* (subtitle of 1996 paperback edition instead: *Rethinking the human place in nature*). W.W. Norton & Co., New York, London.
Cuestas-Caza, J. (2018). Sumak Kawsay is not buen vivir. *Alternautas*, 5, 51–66.
Danowski, D. and Viveiros de Castro, E. (2017). *The ends of the world*. Polity Press, Cambridge.
Davis, M., Chew, M.K., Hobbs, R.J. et al. (2011). Don't judge species on their origins. *Nature*, 474, 153–154.
De Jonge, C. and Whiteman, G. (2014). Arne Naess (1912–2009). In: *Oxford handbook of process philosophy and organization studies* (eds. Helin, J., Hernes, T., Hjorth, D. et al.). Oxford University Press, Oxford, pp. 432–451.
Demeritt, D. (2002). What is the 'social construction of nature'? A typology and sympathetic critique. *Progress in Human Geography*, 26, 767–790.
Descola, P. (1996). Constructing natures. Symbolic ecology and social practice. In: *Nature and society: anthropological perspectives* (eds. Descola, P. and Pálsson, G.). Routledge, London, pp. 82–102.
Descola, P. (2013). *The ecology of others*. Prickly Paradigm Press, Chicago.
Descola, P. and Pálsson, G. (eds.) (1996). *Nature and society: anthropological perspectives.* Routledge, London.
Díaz, S., Demissew, S., Carabias, J. et al. (2015a). The IPBES Conceptual Framework – connecting nature and people. *Current Opinion in Environmental Sustainability*, 14, 1–16.
Díaz, S., Demissew, S., Joly, C. et al. (2015b). A Rosetta Stone for nature's benefits to people. *PLOS Biology*, 13(1): e1002040. https://doi.org/10.1371/journal.pbio.1002040
Dove, M.R., Campos, M.T., Mathews, A.S. et al. (2003). The global mobilization of environmental concepts: re-thinking the Western/non-Western divide. In: *Nature across*

cultures: views of nature and the environment in non-Western cultures (ed. Selin, H.). Springer, Dordrecht, pp. 19–46.

Droz, L., Chen, H.M., Chu, H.T. et al. (2022). Exploring the diversity of conceptualizations of nature in East and South-East Asia. *Humanities and Social Sciences Communications*, 9, 186. DOI:10.1057/s41599-022-01186-5

Ducarme, F. and Couvet, D. (2020). What does 'nature' mean? *Palgrave Communications*, 6, 14.

Egerton, F.N. (1973). Changing concepts of the balance of nature. *Quarterly Review of Biology*, 48, 322–350.

Eser, U. (1999). *Der Naturschutz und das Fremde. Ökologische und normative Grundlagen der Umweltethik.* Campus, Frankfurt.

Flannery, T.F. (1994). *The future eaters: an ecological history of the Australasian lands and people.* Grove Press, New York.

Foreman, D. (1996/1997). All kinds of wilderness foes. *Wild Earth*, 6, 1–4.

Gandy, M. (1996). Crumbling land: the postmodernity debate and the analysis of environmental problems. *Progress in Human Geography*, 20, 23–40.

Glacken, C.J. (1967). *Traces on the Rhodian shore. Nature and culture in western thought from ancient times to the end of the eighteenth century*. University of California Press, Berkeley.

Gordon, G. (2018). Environmental personhood. *Columbia Journal of Environmental Law*, 43, 49–91.

Gould, R.K., Pai, M., Muraca, B. et al. (2019). He 'ike 'ana ia i ka pono (it is a recognizing of the right thing): how one indigenous worldview informs relational values and social values. *Sustainability Science*, 14, 1213–1232.

Groh, D. (2003). *Schöpfung im Widerspruch. Deutungen der Natur und des Menschen von der Genesis bis zur Reformation.* Suhrkamp, Frankfurt am Main.

Gudynas, E. (2011). Buen vivir: today's tomorrow. *Development*, 54, 441–447.

Hacking, I. (1999). *The social construction of what?* Harvard University Press, Cambridge/Massachusetts.

Hager, F.-P., Gregory, T., Maierù, A. et al. (2014). Natur. In: *Historisches Wörterbuch der Philosophie* (ed. Ritter, J.). Schwabe Verlag, Basel, Vol. 6, pp. 421–478.

Haila, Y. (2000). Beyond the nature-culture dualism. *Biology and Philosophy*, 15, 155–175.

Heger, T. and Trepl, L. (2003). Predicting biological invasions. *Biological Invasions*, 5, 313–321.

Heink, U. and Jax, K. (2014). Framing biodiversity – the case of "invasive alien species". In: *Concepts and values in biodiversity* (eds. Lanzerath, D. and Friele, M.). Routledge, London, pp. 73–98.

Helander-Renvall, E. (2010). Animism, personhood and the nature of reality: Sami perspectives. *Polar Record*, 46, 44–56.

Hettinger, N. (2001). Exotic species, naturalisation and biological nativism. *Environmental Values*, 10, 193–224.

Hunter, M.J. (1996). Benchmarks for managing ecosystems: are human activities natural? *Conservation Biology*, 10, 695–697.

Ingold, T. (2000). *The perception of the environment: essays on livelihood, dwelling and skill.* Routledge, London.

Jax, K. (2010). *Ecosystem functioning*. Cambridge University Press, Cambridge.

Jax, K., Calestani, M., Chan, K.M.A. et al. (2018). Caring for nature matters: a relational approach for understanding nature's contributions to human well-being. *Current Opinion in Environmental Sustainability*, 35, 22–29.

Jax, K., Jones, C.G. and Pickett, S.T.A. (1998). The self-identity of ecological units. *Oikos*, 82, 253–264.

Jenkins, W. (2008). *Ecologies of grace: environmental ethics and Christian theology.* Oxford University Press, New York.

Jiao, Y., Ding, Y., Zha, Z. et al. (2019). Crises of biodiversity and ecosystem services in Satoyama landscape of Japan: a review on the role of management. *Sustainability*, 11, 454.

Kalland, A. (2003). Environmentalism and images of the other. In: *Nature across cultures: views of nature and the environment in non-Western cultures* (ed. Selin, H.). Springer, Dordrecht, pp. 1–17.

Kidner, D.W. (2000). Fabricating nature: a critique of the social construction of nature. *Environmental Ethics*, 22, 339–357.

Kimmerer, R.W. (2022). The serviceberry: an economy of abundance. *Emergence Magazine*, October 26. https://emergencemagazine.org/essay/the-serviceberry/

Kirchhoff, T. (2023). (De)Sacralizations of nature in modern Western societies. In: *Multiple sacralities: rethinking the sacred in modern European history* (eds. Gissibl, B. and Hofmann, A.). Vandenhoeck & Ruprecht, Göttingen (in print).

Kitcher, P. (2001). *Science, truth, and democracy.* Oxford University Press, New York, Oxford.

Knauß, S. (2018). Conceptualizing human stewardship in the Anthropocene: the rights of nature in Ecuador, New Zealand and India. *Journal of Agricultural and Environmental Ethics*, 31, 703–722.

Knight, C. (2004). Veneration or destruction? Japanese ambivalence towards nature, with special reference to nature conservation. Thesis, *Master of Arts in Japanese.* University of Canterbury, Canterbury/New Zealand.

Knight, C. (2010). The discourse of "encultured nature" in Japan: the concept of satoyama and its role in 21st-century nature conservation. *Asian Studies Review*, 34, 421–441.

Kohen, J.L. (2003). Knowing country: indigenous Australians and the land. In: *Nature across cultures: views of nature and the environment in non-Western cultures* (ed. Selin, H.). Springer, Dordrecht, pp. 229–243.

Kopnina, H. (2020). Human/environment dichotomy. In: *The International encyclopedia of anthropology* (ed. Callan, H.). John Wiley & Sons, Hoboken, pp. 1–8.

Larson, G.J. (1987). "Conceptual resources" in South Asia for environmental ethics" or the fly is still alive and well in the bottle. *Philosophy East and West*, 37, 150–159.

Latour, B. (1999). *Pandora's hope: essays on the reality of science studies.* Harvard University Press, Cambridge.

Lease, G. (1995). Introduction: nature under fire. In: *Reinventing nature? Responses to postmodern deconstruction* (eds. Soulé, M.E. and Lease, G.). Island Press, Washington D.C., pp. 3–15.

Leeming, D. and Leeming, M. (1994). *A dictionary of creation myths.* Oxford University Press, New York, Oxford.

Lenders, R.H.J. (2006). Ecologist's vision of nature and their consequences from nature valuation and conservation. In: *Visions of nature: a scientific exploration of people's implicit philosophies regarding nature in Germany, the Netherlands and the United Kingdom* (eds. van den Born, R.J.G., de Groot, W.T. and Lenders, R.H.J.). LIT-Verlag, Berlin, pp. 193–210.

Leopold, A. (1949). *A Sand County almanac and sketches here and there.* Oxford University Press, New York.

LeVasseur, T. and Peterson, A.L. (eds.) (2017). *Religion and ecological crisis. The "Lynn White thesis" at fifty.* Routledge, New York, London.

Löfmarck, E. and Lidskog, R. (2017). Bumping against the boundary: IPBES and the knowledge divide. *Environmental Science & Policy*, 69, 22–28.

Lorimer, J. (2015). *Wildlife in the Anthropocene: conservation after nature.* University of Minnesota Press, Minneapolis, London.

Maier, D.S. and Feest, A. (2016). The IPBES Conceptual Framework: an unhelpful start. *Journal of Agricultural and Environmental Ethics*, 29, 327–347.

Mamani-Bernabé, V. (2015). Spirituality and the Pachamama in the Andean Aymara worldview. In: *Earth stewardship: linking ecology and ethics in theory and practice* (eds. Rozzi, R., Chapin, F.S.I., Callicott, J.B. et al.). Springer, Heidelberg, pp. 65–76.

Mann, C.C. (2005). *1491: new revelations of the Americas before Columbus.* 2nd ed. Vintage Books, New York.

Milton, K. (1996). *Environmentalism and cultural theory: exploring the role of anthropology in environmental discourse.* Routledge, London.

Morris-Suzuki, T. (1991). Concepts of nature and technology in pre-industrial Japan. *East Asian History*, 81–97.

Narayan, D., Chambers, R., Shah, M.K. et al. (2000). *Voices of the poor: crying out for change.* Oxford University Press, New York.

Neale, M. (ed.) (2017). *Songlines: Tracking the seven sisters.* National Museum of Australia Press, Canberra.

Northcott, M.S. (1996). *The environment and Christian ethics.* Cambridge University Press, Cambridge.

Nunn, P.D. (2020). In anticipation of extirpation. How ancient peoples rationalized and responded to postglacial sea level rise. *Environmental Humanities*, 12, 113–131.

Nussbaum, M. (2011). *Creating capabilities: the human development approach.* Belknap Press of Harvard University Press, Cambridge.

O'Neill, J., Holland, A. and Light, A. (2008). *Environmental values.* Routledge, London.

Ojeda, J., Salomon, A.K., Rowe, J.K. et al. (2022). Reciprocal contributions between people and nature: a conceptual intervention. *BioScience*, 72, 952–962.

Olenin, S., Minchin, D. and Daunys, D. (2007). Assessment of biopollution in aquatic ecosystems. *Marine Pollution Bulletin*, 55, 379–394.

Page, J. and Pelizzon, A. (2022). Of rivers, law and justice in the Anthropocene. *The Geographical Journal*, https://doi.org/10.1111/geoj.12442.

Peterson, A.L. (2001). *Being human: ethics, environment and our place in the world.* University of California Press, Berkeley, Los Angeles.

Pickering, A. (1995). *The mangle of practice: time, agency, and science.* University of Chicago Press, Chicago, London.

Pickett, S.T.A., Parker, V.T. and Fiedler, P.L. (1992). The new paradigm in ecology: implications for conservation biology above the species level. In: *Conservation biology: the theory and practice of conservation, preservation and management* (eds. Fiedler, P.L. and Jain, S.K.). Chapman & Hall, New York, pp. 65–88.

Plurinational State of Bolivia (2013). Intergovernmental Panel on Biodiversity and Ecosystem Services (IPBES). Submission by the Plurinational State of Bolivia. Conceptual framework for the Intergovernmental Science-Policy Platform on Biodiversity and Ecosystem Services.

Proctor, J.D. (1998). The social construction of nature: relativist accusations, pragmatist and critical realist responses. *Annals of the Association of American Geographers*, 88, 352–376.

Pyne, S.J. (1997). *World fire: the culture of fire on Earth*. University of Washington Press, Seattle.

Ridder, B. (2007). The naturalness versus wildness debate: ambiguity, inconsistency, and unattainable objectivity. *Restoration Ecology*, 15, 8–12.

Rolston, H.I. (1990). Biology and philosophy in Yellowstone. *Biology and Philosophy*, 5, 241–258.

Sandbrook, C. (2015). What is conservation? *Oryx*, 49, 565–566.

Schweitzer, A. (1923/1990). *Kultur und Ethik*. Beck, München.

Selin, H. (ed.) (2003a). *Nature across cultures: views of nature and the environment in non-Western cultures*. Springer, Dordrecht.

Selin, H. (2003b). Introduction. In: *Nature across cultures: views of nature and the environment in non-Western cultures* (ed. Selin, H.). Springer, Dordrecht, pp. xix–xxiii.

Simberloff, D. (2011). Non-natives: 141 scientists object. *Nature*, 475, 36–36.

Simberloff, D. (2012). Nature, natives, nativism, and management: worldviews underlying controversies in invasion biology. *Environmental Ethics*, 34, 5–25.

Simberloff, D. (2014). The "balance of nature" – evolution of a panchreston. *PLOS Biology*, 12, e1001963.

Soper, K. (1995). *What is nature?* Blackwell, Oxford.

Soulé, M.E. (1995). The social siege of nature. In: *Reinventing nature? Responses to postmodern deconstruction* (eds. Soulé, M.E. and Lease, G.). Island Press, Washington D.C., pp. 137–170.

Soulé, M.E. and Lease, G. (eds.) (1995). *Reinventing nature? Responses to postmodern deconstruction*. Island Press, Washington D.C.

Spaemann, R. (1973). Natur. In: *Handbuch philosophischer Grundbegriffe* (eds. Krings, H., Baumgartner, H.M. and Wild, C.). Kösel, München, pp. 956–969.

Strang, V. (2005). Knowing me, knowing you: aboriginal and European concepts of nature as self and other. *Worldviews*, 9, 25–56.

Takeuchi, K. (2010). Rebuilding the relationship between people and nature: the Satoyama Initiative. *Ecological Research*, 25, 891–897.

Takeuchi, K., Brown, R.D., Washitani, I. et al. (eds.) (2003). *Satoyama: the traditional rural landscape of Japan*. Springer, Tokyo.

Taylor, P.W. (1986/2011). *Respect for nature: a theory of environmental ethics*. Princeton University Press, Princeton.

Tengö, M., Hill, R., Malmer, P. et al. (2017). Weaving knowledge systems in IPBES, CBD and beyond – lessons learned for sustainability. *Current Opinion in Environmental Sustainability*, 26–27, 17–25.

Trigger, D.S. (2008). Indigeneity, ferality, and what belongs in the Australian Bush: aboriginal responses to 'Introduced' animals and plants in a settler-descendant society. *The Journal of the Royal Anthropological Institute*, 14, 628–646.

Tucker, J.A. (2003). Japanese views of nature and the environment. In: *Nature across cultures: views of nature and the environment in non-Western cultures* (ed. Selin, H.). Springer, Dordrecht, pp. 161–183.

Turnbull, D. (1997). Reframing science and other local knowledge traditions. *Futures*, 29, 551–562.

UNEP (2014). Conceptual framework for the Intergovernmental Science-Policy Platform on Biodiversity and Ecosystem Services. Decision IPBES-2/4. In: Report of the second session of the Plenary of the Intergovernmental Science-Policy Platform on Biodiversity

and Ecosystem Services. www.ipbes.net/sites/default/files/downloads/Decision%20IPBES_2_4.pdf
UNESCO (2013). The contribution of indigenous and local knowledge systems to IPBES: Building synergies with science. UNESCO, Paris.
Viveiros de Castro, E. (1998). Cosmological deixis and Amerindian perspectivism. *The Journal of the Royal Anthropological Institute*, 4, 469–488.
von Droste, B., Plachter, H. and Rössler, M. (eds.) (1995). *Cultural landscapes of universal value*. Gustav Fischer, Jena, Stuttgart, New York.
Wallington, T.J., Hobbs, R.J. and Moore, S.A. (2005). Implications of current ecological thinking for biodiversity conservation: a review of the salient issues. *Ecology and Society*, 10(1), 15. www.ecologyandsociety.org/vol10/iss1/art15/.
Warren, K.J. (2015). Feminist environmental philosophy. In: *The Stanford Encyclopedia of Philosophy* (Summer edition) (ed. Zalta, E.N.). http://plato.stanford.edu/archives/sum2013/entries/well-being/
Weiss, G. (1972). Campa cosmology. *Ethnology*, 11, 157.
White, L.J. (1967). The historical roots of our ecological crisis. *Science*, 155, 1203–1207.
Williams, R. (1980). Ideas of nature. In: *Problems in materialism and culture* (ed. Williams, R.). Verso, London, pp. 67–85.
Woods, M. and Moriarty, P.V. (2001). Strangers in a strange land: the problem of exotic species. *Environmental Values*, 10, 163–191.
Zent, E. (2015). Unfurling western notions of nature and Amerindian alternatives. *Ethics in Science and Environmental Politics*, 15, 1–19.

5
MOVING FORWARD
Which conservation concepts for which purposes?

The previous chapters have been largely analytical. In Chapter 3 I have elaborated on the various conservation concepts and arranged them into a number of clusters. Even though some clear distinctions and potential conflicting approaches between these major concepts are evident, a closer look reveals that sharp delimitations in terms of the objects and values embraced are often much more difficult to make than it appears at first sight, given the various interpretations of major concepts such as biodiversity, ecosystem services, and even wilderness. As an alternative to sorting the concepts along values and objects alone, I suggested to approach them by a narrative analysis of the human–nature relationships implicit in the concepts. Also here a completely clear-cut one-to-one attribution between specific narratives and specific conservation concepts is not always possible, but the relations are less ambiguous than when ordering concepts by values or objects alone. I think that approaching conservation concepts via their underlying human–nature relationships is a useful path; narrative descriptions of human–nature relationships *include* values and objects of conservation but at the same time are less abstract and closer to human actions. This also links to the fact that, as discussed in Chapter 4, human–nature relationships are at the heart of both ideas of what (proper) nature is and of our own self-understandings as humans, variable in time and between societies, and even between groups of people within a society.

The current chapter builds on the previous analyses and strives for a more reflexive use of conservation concepts in specific settings. The question with which this book started was: "*How do we want to form (or reform) our relationship with nature as humans – and which conservation approaches can help us?*". Taking up this question, I will, more specifically, ask in this chapter:

> *How can we select the most adequate concepts for particular biophysical and social boundary conditions and find proper human-nature relationships that allow for a sustained conservation of nature and human well-being alike?*

In pursuing this question I have to stress that this chapter is not providing a "manual" for the direct practical application of conservation concepts. What is offered instead is a presentation of criteria for selecting conservation concepts and a short procedural suggestion for clarifying conservation goals and tackling conservation conflicts.

In entering this forward-looking chapter, it is necessary to state and explain more about my own position as regards conservation. I have tried to be as unbiased as possible in the previous chapters; nevertheless it is unavoidable that my own understanding of conservation, nature, and what it means to be human, have influenced my writing to some degree. But it is impossible to be "neutral" when it comes to making recommendations. So let me start with some premises on which the following builds.

A first important premise that guides this chapter is that there is not one conservation concept and strategy that fits all purposes – in the sense that it can be ideal for all people and locations, for all societal and biophysical contexts.

Another premise is already prominent in the phrasing of the above question, namely that good (or proper) human–nature relationships should account both for the well-being of humans and non-human nature. As I will discuss later, it is not the least the very relations as such that count as something to be valued.

Also, related to the former, I presume that, as a result of the elaborations in previous chapters, that human–nature relationships – via specific ideas about nature – are linked to peoples' self-understanding as humans.

When it comes to implementing conservation, however, yet another dimension comes in more explicitly, which features prominently in this chapter. We cannot separate our self-understanding of what it means to be human, as it is reflected in our relations to nature, from that of our relations to other human beings. Both belong together.

Moreover, like I do not consider any one concept being the solution to all conservation problems, I also do not see that it is necessary to commit oneself to one type of human–nature relationship (e.g. in the sense of the above narratives) only. It is undeniable that all humans depend on nature in basic ways. The "Safeguarding necessary and useful nature" narrative is thus not a bad thing for conservationists but something we *have* to share, out of simple biological needs. But I am convinced, as for myself and as what I perceive as at the heart of most conservationists' actions, that we also need other types of relations in addition, relations which open up to other living beings, human and non-human alike, in an attitude of respect and care. For me, this is something that is part of what it means to be really human.

I will further develop and justify these premises below. To do so, I will first take a look back and ask if, and in what ways, we may see real progress in the succession of conservation concepts over time, from the beginnings of organised conservation in the 19th century until today – and what that might mean for a pluralistic approach towards conservation (Section 5.1). Section 5.2 will then further broaden the perspective, beyond ideas of human–nature relationships towards including also human–human relationships. This is necessary both on philosophical grounds as on practical ones, because conservation does not take place in a societal vacuum but is embedded into social and economic conditions, raising important questions about power and justice. Especially it questions the common "we" of referring to humanity as a whole with respect to conservation. Based on this and on the results of the previous chapters, I will discuss what I see as major criteria for selecting appropriate conservation concepts in specific settings (Section 5.3). The last section (Section 5.4) will then present a procedural suggestion for clarifying conservation goals and tackling conservation conflicts.

5.1 Conservation concepts: progress or recurrence of always the same?

In a much-cited paper, the late Georgina Mace presented a figure that, at first sight, suggests a kind of succession of conservation concepts as well as of prevailing ideas about human–nature relationships ("framings of conservation") since the 1960s (Figure 5.1). Mace, however, acknowledges that "none of the framings has been eclipsed as the new ones have emerged, resulting in multiple framings in use today" (Mace 2014, p. 1559). I would go even one step further here. If we extend the view beyond those discourses dominant in the (English-language) literature and also through time, one can see that even the "new" framings have a much longer history. In fact most of them have already been present simultaneously almost since the beginnings of conservation in the late 19th century (see Section 2.2), even though often without much scientific underpinning (or none at all). Thus, ideas of what Mace labels "People and nature" have been present in the form of cultural landscape protection early on (e.g. with Ernst Rudorff's "*Heimatschutz*"). The beginnings of a perspective of what today is called social-ecological systems perspective (as one key word of the "People and nature" framing"), with a stronger emphasis on science, can already be seen in the highly influential writings of Aldo Leopold and his idea of the "land" as a community of both humans and non-human beings (Meine 2020). Likewise, we find Mace's "Nature for people" framing already with Gifford Pinchot's forest management or even in the writings of George Perkins Marsh. The "Nature for itself" framing also was there as an idea from the

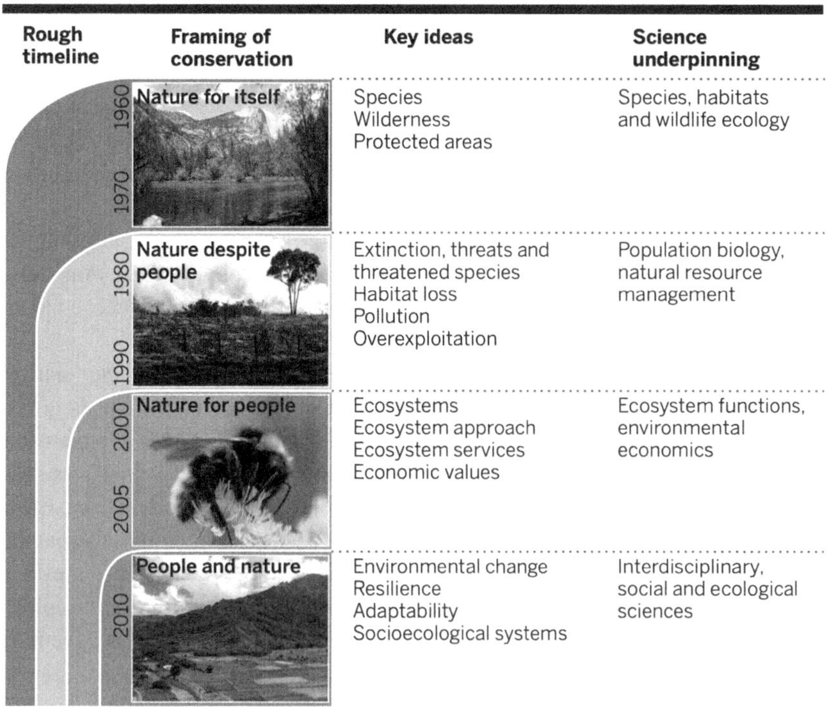

Rough timeline	Framing of conservation	Key ideas	Science underpinning
1960–1970	Nature for itself	Species Wilderness Protected areas	Species, habitats and wildlife ecology
1980–1990	Nature despite people	Extinction, threats and threatened species Habitat loss Pollution Overexploitation	Population biology, natural resource management
2000–2005	Nature for people	Ecosystems Ecosystem approach Ecosystem services Economic values	Ecosystem functions, environmental economics
2010	People and nature	Environmental change Resilience Adaptability Socioecological systems	Interdisciplinary, social and ecological sciences

FIGURE 5.1   Changing views of nature and conservation since ca. 1960.

Source: Reprinted from Mace, G.M. (2014). Whose conservation? *Science*, 345, 1558–1560; Reprinted with permission from AAAS. See text.

beginning of conservation. With respect to this latter framing, I would, however, see Mace's two "key ideas", "wilderness" and "species", as corresponding to two different narratives of human–nature relationships, namely "Nature apart from humans" (wilderness) and "Caring for other living beings is part of a good human life" (species) (see Section 3.3).

Much recurrence but also some important trends

The analyses performed in Chapters 2 and 3 show that there is not a clear succession of conservation arguments during history; most arguments, framings, and narratives[1] have been present since the 19th century but the latest by the mid-20th century. Clearly there have, however, been shifting emphases, and in particular there have been new terms to conceptualise, communicate, and operationalise these arguments.

Nevertheless, some trends can be clearly described and are important for today's debates on conservation concepts. As already noted at the end of Section 2.2, there was a strongly increasing influence of the sciences during the 20th century, at first biology and natural resource management, and soon also ecology. It was only in the 1990s that economics became part of the conservation mainstream, and only in recent years attention to the social sciences, anthropology, also environmental ethics, grew within *mainstream* conservation. With the rise of the natural sciences and economics in conservation, along with improved empirical techniques as well as data handling, especially in ecology, the amount of quantitative data has skyrocketed. This development was very positive on the one hand, creating the basis for a much more reliable description of the actual ecological situation, better ideas about the economic importance of nature to society beyond direct use values, and it also helped in developing practical conservation tools.

But the increasing scientification and quantification (including various quantitative methodologies in ecological economics) has also been in danger of pushing back the traditionally important social, cultural, and symbolic dimensions of conservation to an issue of secondary importance. In part this has led to a more generalising and global perspective on conservation, at the loss, however, of the more concrete, local, and specific. Also, even the best quantification of economic value and ecological "facts" is futile if the theories and concepts it is meant to support are too vague, as is still often the case in conservation biology and ecology.

This brings me to a second trend: since the last two decades of the 20th century, an increase in the creation of concepts that serve as boundary objects can be observed, as well as a highly strategic orientation of "new" conservation concepts. Boundary objects are meant to facilitate communication across disciplinary borders by creating shared vocabulary although the precise understanding of the term may differ between those who use it (see Chapter 1). As already described, many new terms were created to appeal to particular new audiences and to actively promote a conservation agenda outside of the traditional conservation community. In Section 3.4 this has been described in some detail for the concept "ecosystem services". Also for the term "biodiversity", where one might not expect it at first sight, it has been very well-documented how it was created explicitly as a neologism to address policy and the general public. The story has been told quite often already, following David Takacs' (1996) excellent book, but it is worth repeating here (see also Eser 2002). In 1986 a large conference with the title "National Forum of BioDiversity" ("biodiversity" then still with a capital D) was organised in Washington D.C. It was sponsored by two organisations of high scientific reputation, namely the Smithsonian Institution and the US National Academy of Sciences (NAS). The conference sprang from concerns about the accelerating rate of species extinctions and aimed at getting the attention of US politics on this problem. Dan Janzen, one of the speakers on the Forum on BioDiversity told Takacs in an interview:

> The Washington Conference? That was an explicit political event, explicitly designed to make Congress aware of this complexity of species that we're losing. And the word [biodiversity] was coined – well different people get credit

for coining the word – but the point was the word was punched into that system at that point deliberately. A lot of us went to that talk with a political mission. We were asked, will we come and do this thing? So we did.

(Takacs 1996, p. 37)

This happened in spite of major concerns by the NAS that the Forum might be too much an advocatory one instead of being on "objective" science, in line with the reputation of the Academy. Some 14,000 people attended the conference, which Takacs (p. 39) describes as "consciousness-raising event and media spectacle" and just this consciousness-raising was at the heart of what the organisers hoped to achieve.

Also the *term* biodiversity itself was coined to allow for easier public acceptance and proliferation of the ideas behind it, namely the "species extinction crisis". As Walter G. Rosen, one of the organisers of the conference told Takacs:

It was easy to do: all you do is take the "logical" out of "biological".[…] To take the logical out of something that's supposed to be science is a bit of a contradiction in terms, right? And yet, of course, that's why I get impatient with the Academy, because they're always so logical that there seems to be no room for emotion in there, no room for spirit.

(Takacs 1996, p. 37)

So what happened, as also Eser (2002) has argued, is that the new term served two aims: on the one hand, it should appeal emotionally ("diversity is good") to a non-scientific audience by avoiding bulky terminology; at the same time it should still retain its scientific reputation as a technical term, as being "objective". In what followed, scientists tried to sharpen the meaning of the term and used it indeed as a measurable and almost exclusively scientific concept, even though its early protagonists saw "biodiversity" as a shortcut for "nature as a whole" or even "nature conservation" (interviews conducted by David Takacs, Takacs 1996, p. 46ff). This breadth of understanding has persisted until today (see Section 3.4). It is part of the success of biodiversity as a boundary concept that it unites different perspectives and allows many people to agree on biodiversity as a shared goal, even though they might not mean exactly the same by it.

This strategic dimension is likewise very strong with the ecosystem services concept and related terms, here addressing an audience really or seemingly focused on economic issues. Also new terms in the ecosystem services cluster, such as "nature-based solutions" or "green infrastructure", were coined with an impetus to address politics and the economy. And yet, again it was not the least a strategic reason to replace the term "nature's benefits to people" by "nature's contributions to people" (NCPs) in the IPBES framework, here to counter some undesired "side effects". It was inter alia a reaction to critique regarding the economic connotations of "ecosystem services". Strategic moves in terms of a particular targeted wording are easy to make but difficult to reverse once they have been taken up and are established. In some cases, strategic changes in wordings, however, can also be very successful. To avoid negative connotations of the word "reserve", and, related to that, local resistance to the establishment of UNESCO Biosphere Reserves, the term was changed in Austria into "Biosphere Parks", allowing local people to identify themselves more easily with it.

To sum up, most major concepts currently in use embrace values and human–nature relationships which have been around under various names long since. One exception might be the concept of "novel ecosystems" along with the narrative I have called "Gardening nature in novel ways". Here, it is not that areas which are today dubbed "novel ecosystems" were new, but that such areas were only recently conceptualised in the current manner and, what's most important, discussed as legitimate objects of conservation.

A proliferation of conservation concepts: (when) is it a problem?

Why do conservationists coin new terms at all, if much of what is behind these terms has already been expressed earlier by means of other terms, at least conceptually similar ones? What is the purpose to introduce *new* concepts and terms? There can be, in principle, several good reasons for this.

One reason is that described earlier: new terms are coined to reach new audiences. New terms are, however, also coined to remove obstacles related to undesired connotations of old terms. Thus "biodiversity" seemed more appealing to many potential conservation partners than, "endangered species", "nature conservation" or simply "nature", both in the scientific realm as in politics, business, and the general public. "Nature conservation", e.g., was often deemed old-fashioned or related to restricting people's actions, mostly in protected areas.

A more classic and important reason to coin new terms is that they might serve better to describe and classify certain issues than existing ones. "Novel ecosystems" might be such a term, but also, in addition to its strategic impetus, "ecosystem services". The potential benefits of nature to humans were conceptualised not as before, simply under the heading of "natural resources", but by adding and distinguishing contributions such as regulating services and cultural services (e.g. in the MA, see Section 3.3). Also, sometimes a new umbrella term for a larger set of ideas and phenomena may be needed, as with "biodiversity" or "ecosystem functioning". Finally, a new, at first still vague concept may have heuristic value for furthering and developing new ideas, beyond the classical categories of established concepts and theories.

There are thus good reasons for fresh concepts and terms, and I will make a plea below to maintain a plurality of conservation concepts. Nevertheless, there are also problems with too many conservation concepts and too fast a succession of new terms and concepts; this was one of the starting points for writing this book.

The first problem is simply that confusion can arise with too many new terms, both internally (within the conservation community) or for the public. As Section 3.4 demonstrated, distinctions and relations between concepts are often not really clear, partly on behalf of their inherent vagueness, as with biodiversity, partly because multiple meanings are attached to them (as, e.g. to "stewardship", "ecosystem functioning", or "resilience" in a conservation context).

The question on the relation between biodiversity and ecosystem services, e.g., has fostered many discussions, not the least fuelled by the concern that some valued aspects of biodiversity might become lost (or invisible) with a focus on ecosystem services, while others argue that ecosystem services protection begets biodiversity protection (see discussion in Jax and Heink 2015). As another example, the aforementioned partial "replacement" of "ecosystem services" by NCPs has evoked criticism on several levels. One of these is that many practitioners in public administrations and NGOs, who meanwhile had become used to the ecosystem services concept, had a problem with the somewhat subtle distinction between ecosystem services and NCPs adding up to the linguistic bulkiness of the new expression.

Another problem is that if new conservation concepts follow each other too rapidly, research often is shifting too fast to new areas, especially if connected with new funding schemes that may divert many researchers to the seemingly new area. The same goes for the desired public attention, which also takes some time to accept and use new concepts and terms.

I sometimes wonder, if we, as conservationists, are too impatient to follow a few major lines of argument, with its associated research, and major conceptual and strategic approaches. Impatience is understandable to some degree, in the face of nature vanishing more and more, faster and faster, and with the attention span of politicians and the general public often not as long as it should be. Beyond that it is also very difficult to really assess the success or effectiveness of conservation concepts and associated approaches, given that multiple factors influence it and the existence of long time lags between implementing conservation concepts and their effects on conservation targets.

Arguments for a pluralistic approach

The opposite of pursuing a confusing multitude of conservation approaches is focusing on a too restricted set of conservation concepts and the search for panaceas.

I argue in favour of a pluralistic approach towards conservation. In the following I will explain why I think that is necessary and how to find a middle ground between a kind of "monism" and an overly proliferation of concepts.

There are at least three objections against searching for one conservation approach and one human–nature relationship reflecting it only. First, conservation goals are highly diverse, as are the objects of conservation. There is not one way, even with presumably "holistic" approaches to conservation, such as ecosystem management, that can guarantee the protection of all potential goods that conservationists strive for. A hands-off-approach, as some wilderness proponents postulate, often may neither lead to a maximum of biodiversity (e.g. in terms of species or habitat diversity; see Sarkar 1999) nor of ecosystem services. Other trade-offs prevail. Bryan Norton has postulated the convergence of the effects of different conservation values, as eventually leading to the same conservation policies (Norton 1991). Even though Norton may be right that there is no deep rift between anthropocentric and biocentric approaches, and that serving human interests and protecting other species on behalf of their presumed intrinsic value can be reconciled, in practice differences will remain. There is no way to harmonise a strict wilderness idea of nature without humans and that of a cultural landscape in the same place.

Second, there is no "ideal" state of nature, no privileged concept and reference point. There is no evident target state for conservation and restoration: not "the natural" (see Section 4.1), not a particular historical state. Nor is there an evident object of conservation, e.g. species or ecosystems. If there were, things would appear easier (in fact there would still remain societal choices and options, even then), but already in the face of natural dynamics and even more under ongoing global change such ideal benchmarks are an illusion. Different conservation concepts aim at different target states for conservation (see earlier).

Third, the search for panaceas easily leads into muddy social waters. In the extreme it may steer into attempts of creating *purity*. "Pure wilderness" especially, also "nativism" is prone to cause conflicts between conservationists. Sometimes the conflict may also arise not so much from how (e.g.) wilderness protagonists perceive wilderness but how their opponents do (strawmen arguments…). Particular approaches are sometimes seen as the silver bullet for all conservation problems by some groups or as pure poison to conservation by others. Some years ago, I was involved as co-lead in a large project on the operationalisation of the ecosystem services concept. During a panel discussion close to the end of the – in general very open-minded – project I tried to articulate some limitations of ecosystem services approaches as I perceived them. My statement evoked strong objections from other participants on doubting the all-embracing potential of the ecosystem services approach. A few weeks earlier, in contrast, presenting in a small workshop of conservation-minded ecologists, I had to defend my careful appraisal of the ecosystem services concept against the majority opinion that saw the ecosystem services concept as something that endangered "real" conservation and should be avoided at all costs.

In the extreme, the search for purity and unambiguity is something that may lead to dogmatism, its proponents seemingly being in possession of the "truth" (see Douglas 1966/2002, Bauer 2018). This is not in the first place a problem of conservation but even more in other societal fields such as with well-known forms of nationalism, racism, or religious fundamentalism.

A plurality of conservation approaches instead allows to adapt conservation to different social and biophysical boundary conditions and it can also account for previously neglected and minority positions, within society and within the broader conservation community (Matulis and Moyer 2017). The tension between different concepts can be productive, if it is acknowledged that there is not one exclusive true and right conservation approach but that the task is to find *appropriate* ones for the tasks and settings at hand. It can and should be a basis of negotiations. Nevertheless, it is obvious that not all conservation ideas, not all ideas of proper human–nature relationships can be realised or accounted for in one and the same place. Choices are necessary and conflicts may arise, as well as winners and losers resulting from particular conservation practices (see Sections 5.2 and 5.3).

BOX 5.1   A PERSONAL EXPERIENCE

In my personal experiences, I cannot subscribe to one idea of valuable nature either. Growing up in the cultural landscapes of the Rhine valley and the Eifel highlands in Germany, I learned to cherish these landscapes: as home, as an aesthetically pleasing surrounding, as the habitats of species, as a place to experience solitude. It was and is no doubt for me "nature" worth protecting. Nevertheless, I have also, early on, been fascinated by the wilderness idea and still find it wonderful and good that such areas, where humans interfere as little as possible, exist. But one of my strongest experiences with wilderness was quite unexpected, took me by surprise, and changed my perspective on it. In 2002 I was leading a student group from Munich in an excursion to the Chilean island of Navarino, an island of the Tierra del Fuego archipelago, south of the Beagle Channel. I had been there before already twice, on invitation of my Chilean friend and colleague Ricardo Rozzi. This time, part of the programme we had developed was a long hike from the North Coast to the legendary site of Wulaia on Navarino's western coast. Wulaia is the place where – inter alia – Darwin and Fitzroy met the native Yaghan people and from where Fitzroy took some of them to England, most famously Jemmy Button. But also other early encounters between the British and the native Yaghans occurred there, in part also violent ones at the cost of the Europeans. The site is extremely remote, accessible only by boat or on foot. So packed with way too much baggage (including tents) we started in the early morning walking through the hills covered with almost "pristine" *Nothofagus* forests. The area has been used by the native Yaghan people only very extensively. Being water nomads, their main places to dwell were the coastline and the sea. So this was as much of a wilderness as I could imagine. It was damp, much dead and rotting wood everywhere, there was no real trail, and at times I was not sure if even our local guide knew himself where we were. No connection to the human world at all. The hike turned out to be much longer than anticipated. We were supposed to reach Wulaia that same evening but in the end we were lucky to make it at least to the western shore, to what turned out to be Bahia Inutil, the "Useless Bay". Everybody was happy that we now at least had a better idea of where we were. The hike up to this time had been very strenuous and even dangerous in parts, but also highly fascinating, as was the evening at the campfire. The next day, it took another four hours along the coast to reach Wulaia. Wulaia (Figure 5.2) is a magic place, with an amazing location, some small rocky islets sheltering the cove, but it is not really a beautiful place today: an old, abandoned postal and military stone building, surrounding forest and some meadows in rather degraded state (due to some free-ranging cattle and occasional firewood collection), no people. And yet it struck me completely unexpectedly when I first had a look on Wulaia that day. My feeling was that the beauty of the wilderness was only *complete* with these tiny traces of humans in the landscape and its human history. Wilderness was beautiful, it was at the same time ambivalent, and even hostile and dangerous (often part of its fascination). But the small human traces made it even more what it was and what made it even more valuable to me. A strange kind of dialectics.

FIGURE 5.2 Caleta Wulaia (Wulaia cove), Isla Navarino, Chile. View from the north. Photo: Kurt Jax.

A last point why a pluralism of conservation concepts is necessary is that – for the most part – conservation is not a global but a local issue. Striving for generalities is important, but, like traditional and indigenous knowledge and like ecology (see Trepl 1987, Simberloff 2004), conservation is not only a matter of general principles or even laws but is strongly devoted to the specific, to the particular, to places. Each place is different, means different things to people, and also its history matters. Conservation approaches should account for these differences, these histories.

5.2 The social and economic background of implementing conservation goals

Conservation is one among many human activities. It is performed by people; it affects people in many different ways, both positive and negative; it takes place in very concrete social, cultural, and economic settings. Within these settings, different groups of people have different access to resources, have different influence on/in decision processes, and are affected in different ways by decisions. Such effects concern direct economic interests and issues of rights (e.g. property rights as well as access to particular areas and resources) but also non-material ones, touching upon questions of respect and self-understanding of people as related to nature, including spiritual dimensions. On behalf of this, conservation has also to deal with issues of power, justice, and mutual respect.

What is justice? Being a moral term, justice has several related meanings (Box 5.2). As an overall characterisation, justice is about the appropriate treatment of others, and even more about human obligations to others. As such it describes "what we owe to others: to our fellow human beings today, to future generations or to non-human entities" and thus "refers to actions that we have reason to demand from others" (Eser et al. 2014, p. 89f). Power, then, is the ability to influence or control what others do or think (MacMillan online Dictionary, modified). Power is exerted, for good or for bad, on other humans and on non-human beings.

Quite often, conservation is depicted as powerless and in a defensive position, fighting against major economic and political interests. This is true, and it is a major difficulty for implementing conservation in an effective way. However, there have been and are many cases where the reverse is the case (Chan and Satterfield 2013, Sandbrook 2017), i.e. where conservation exerts power onto people. The latter comes with several faces: some quite obvious, some more hidden, even to conservationists themselves.

Power, justice, and the colonial heritage of conservation

Section 2.2 (especially Box 2.1) already dealt with the colonial roots of conservation. In many early protected areas native people have been evicted for the sake of creating "pure" nature, be it in colonial Africa or Asia, be it in US National Parks like Yosemite and Yellowstone. But even in later times and until today the practice continued (see, e.g. Dowie 2009, Agrawal and Redford 2009 for overviews), giving rise to critique of conservation, especially as conducted by some powerful international NGOs, as being neo-colonialist. This refers, in particular, to the establishment of large wilderness areas and other "people-free" conservation areas, a practice also dubbed "fortress conservation" by its opponents (Hutton et al. 2005, Brockington and Igoe 2006). Within the social sciences, this has even caused a hostile attitude towards conservation at large by some authors, even though, in

BOX 5.2 JUSTICE AND POWER

Justice is about the appropriate and fair treatment of others, and about human obligations to others (Chan and Satterfield 2013, Eser et al. 2014). Justice does not always mean equal treatment. Justice for animals does not mean that they have exactly the same rights as humans or that they are treated in exactly the same way. Justice is dependent, inter alia, on specific needs and capabilities. Also for justice among humans, justice does not mean that everybody must be treated completely equally; e.g. children are not treated the same as adults when it comes to voting rights, but it is deemed unjust in modern democracies to deny voting to adults based, e.g., on gender or economic status. There are various theories of how to determine and achieve justice, theories whose treatment is beyond the scope this book (overviews can be found, e.g. in Miller 2021; specifically for biodiversity and for animals, respectively, see Chan and Satterfield 2013, and Nussbaum 2022).

In a conservation context, a distinction is sometimes made between environmental justice and ecological justice (Eser et al. 2014, Shoreman-Ouimet and Kopnina 2015). *Environmental justice* here refers to justice between groups of people who are affected differently by human activities in connection with nature or the environment, such as when the use of natural resources affects especially the poor and profits especially the wealthy, or when indigenous people are removed from conservation areas. *Ecological justice* refers to justice between human and non-human beings, such as when habitats of non-human organisms are destroyed for building a new entertainment facility. Both types of justice do not necessarily exclude each other.

Justice, especially between humans, is not only relevant as *distributional justice*, i.e. the fair distribution of benefits or disadvantages ("costs") deriving from human actions. What can be as important is *procedural justice*, meaning, e.g., that people are equally well involved in decision processes or are brought to a similar level of knowledge about the issues at stake (e.g. Paloniemi et al. 2015).

Power, as "ability or capacity to either act oneself or direct the action of others" (Rogers et al. 2013, entry "Power") can infringe on justice but it can also promote and enforce it. Power relations, especially those that lead to injustices between groups of humans, are easily overlooked. They even occur within the community of conservation scientists and conservationists, between, e.g., mainstream opinions and marginalised ones. Power can be enacted directly by individual people or more subtle as structural power, engrained in established societal and economic structures and processes, even by acting on what is taken for granted for some but may not be for others (e.g. specific ideas about "nature"; see Chapter 4).

fact, the situation is more complex as regards the effects of conservation on human livelihoods (see Shoreman-Ouimet and Kopnina 2015 and later).

Political scientist Arun Agrawal (1997) describes structural similarities between the discourses on colonial development, postcolonial development, and modern conservation in tropical or, more general, "developing" countries. A central persistent theme he exposes is the (Western) idea of *progress*, going along with specific expressions of ("Western superior") civilisation and the superiority of Western knowledge systems (i.e. science). Progress, a difficult concept with many ambiguities (see Box 3.2), at first served the development of the colonial powers themselves, with the colonies providing resources for this development, and for bringing "civilisation" to the "undeveloped" people of those countries. In the postcolonial development discourse, progress meant in fact to bring these "undeveloped" countries themselves to the level of the Western "developed" countries, always mainly by means of advanced and – at least implicitly deemed – superior Western knowledge and technology. Conservation then, for Agrawal, is in continuity with these mindsets and ideas of progress and development:

> The progressive mission that Western colonizers had to fulfill to develop the forces of production in the colonies now is displaced on to the need to conserve natural resources. And now, once again, resources that require conservation are located in the tropical countries of the Third World; these resources constitute mankind's patrimony – so the argument goes – and are being squandered by the countries that possess them; the only hope for conservation lies in massive research and transfers of technological expertise to the countries that possess biological resources.
>
> *(Agrawal 1997, p. 471f)*

Explicitly without meaning to discredit either modern development efforts or the need for conservation, Agrawal's analysis points at some important concepts which turn out to be double-edged swords. The argument of "civilisation" has been used for good as well as for bad in the history of conservation. It has been used for good by forbidding particular sports-hunting practices in the 19th century, the prevention of cruelty towards animals as a duty of civilised humans, but also for bad, e.g., when traditional behaviours such as hunting practices of indigenous peoples in Africa were prohibited (and people removed from their areas) under the label of civilisation, even though "civilised", apparently less cruel trophy hunting by the European colonial rulers was still perpetuated. This went hand in hand with the then common idea that indigenous people were inferior to white people from Western civilisations. So if today national parks and other areas for nature protection are established under the label of being a duty of civilisation and a common duty of mankind ("mankind's patrimony") reflections on the specific social conditions of the respective countries and regions are necessary. Power today is not exerted by external colonial states any more, but more by the postcolonial states themselves or by large international NGOs. Power, also, does not need to be brute force in the first place, although even today local people are sometimes evicted from their home places or restricted severely in their ways of living when new protected areas are established for conservation (see also recent trends of a militarisation of conservation in some countries, e.g. Duffy 2014). Other, more subtle and less conscious modes of wielding power are given by the hegemony of ideas and values produced by Western societies and by science, by the power of defining discourses and setting agendas, and by economic inequalities (Paulson et al. 2012). Progress, in spite of internal critique, is still largely taken for granted as being a good thing in Western societies, but it may not always be that to local people, either because they have other ideas of what would constitute progress to them, do not see progress as good in the first place, or do not have a concept of progress at all. Likewise, many conservation concepts, even the concepts of "nature" and the "natural" that originated in the West are rather alien to other cultures (see Section 4.3) and have simply been imposed on them. This happens in the course of the overall dominance of the Western economic, social, and scientific thinking and – unconsciously – by taking one's own ideas for granted or even to be human "universals", common to all people on Earth. The difficulties of putting Western science, its concepts and its methods, on par with other knowledge systems has already been discussed in Chapter 4. Even the seemingly indisputably good idea of nature (or biodiversity) as the "patrimony of humankind" can be problematic in this context (see Agrawal's quote earlier).

It is a matter of justice, to respect other cultural approaches to nature, which must move us (as conservationists) to open up our understanding of and our approaches to nature conservation, thereby refraining from global solutions and moving to regionally adapted ones. Moreover, as ideas of nature, can be highly diverse even within a small area (see, e.g. Berghöfer et al. 2010), the challenge is then how to deal with such diverse ideas in a fair manner.

In the extreme, concerns for human livelihoods can be pitted against that for non-human nature. In fact, the implementation of conservation does not always produce win-win solutions – neither between different conservation interests nor between conservation and other interests (McShane et al. 2011, Muradian et al. 2013). There may be trade-offs, with winners and losers. But that does not mean that conservation and human livelihoods are irreconcilable. There are many conservation narratives and approaches which aim at harmonising them, even though not always in one and the same place.

Issues of power and justice thus go far beyond the subject of establishing protected areas. (Unintended) consequences of specific conservation approaches also surface, e.g., with the use of the ecosystem services concept. Jax et al. (2013) describe a number of ethical issues related to this concept. Three aspects can be highlighted in particular: (1) who makes the choices regarding the use of ecosystem services?; (2) which values and which objects of nature are included or highlighted and which are excluded or obscured?; and (3) who is impacted (positively or

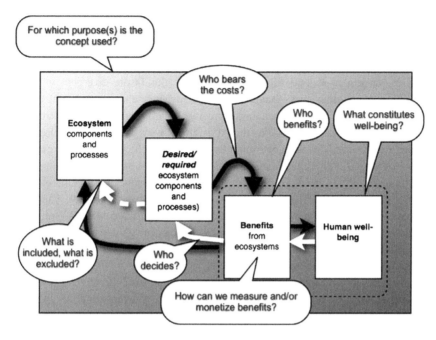

FIGURE 5.3 The generic idea of ecosystem services and a number of ethically relevant questions related to the ecosystem services concept. Black arrows depict causal relations between ecosystems and human well-being. What counts as service is subject to societal choices (white arrows) about what benefits are and which ecosystem processes and components are considered as desirable to promote these. Note that in some common definitions the "desired/required ecosystem components and processes" are called "ecosystem services" in the narrow (measurable) sense, in others the "benefits derived from ecosystems". Bubbles indicate ethically relevant questions; see text.

Source: Figure reprinted from Jax, K., Barton, D.N., Chan, K.M.A. et al. (2013). Ecosystem services and ethics. *Ecological Economics*, 93, 260–268, with permission of Elsevier.

negatively) by choices regarding ecosystem service use (see Figure 5.3)? All three issues are, however, not restricted to the ecosystem services concept, but to different degrees, also apply to other conservation concepts.

The first aspect described is again about the *power of definition and framing*: who has a voice that can be heard when defining, selecting, and managing ecosystem services, and therefore who can make a difference? While the concept has become more relevant for policies and management decisions, the voices of those who benefit "on the ground", those who affect ecosystem services or are affected by their use, are less often heard. Selecting what are relevant ecosystem services is linked closely to the question of what constitutes human well-being, which is not automatically the same for all groups of people. If values and aspects of human well-being that are important to some people are not included in the discussion, features of nature that contribute to these values may be neglected, with the further consequence of overlooking entire modes of living and disparate understandings of what constitutes a good life. An example for this is the divergent views of the official Australian conceptualisation of well-being and those of Australia's Aboriginal people. Kinship to natural and mystic objects and access to sacred places, for instance, are crucial to the Aborigines' way of life (see Section 4.3), but are not covered by the government's definition, which focuses on classical Western ideas of well-being (Sangha et al. 2015). While some basic components of human well-being (such as food and shelter) are the same for all humans, many others, which are culturally and environmentally determined, are not. There is a danger of (unwittingly) imposing a simplified and restricted idea of well-being on all humans – based again on mainly Western ideas as guided by contemporary science and the economic mainstream.

The second aspect is related to the first, here in terms of the contents of the choices. The question here is that of determining the *kind of values and objects that are covered* by the ecosystem services concept, and those that are possibly neglected and consequently impaired by using it. This is an issue that poses risks both to non-human beings as well as to the ideas of a good relationship with nature embraced by people. Excluding some aspects of nature from an ecosystem services framework will thus put these aspects at risk when decisions are taken on how to deal with nature. What is not perceived as an ecosystem service may not be considered as being of value, and thus neglected. If the focus is, e.g., just on species considered useful for the provision of ecosystem services, those species who are deemed not useful or even detrimental will be eclipsed, as will be the values and interests of those people for which these species are important for other reasons (e.g. being sacred).

As said earlier, this is not unique to the ecosystem services concept. Already the selection of a particular conservation concept is a matter of choice (see, e.g. case study on the Oostvaardersplassen in Section 3.1) and also within the major concepts choices have to be made again, e.g. what is the specific aspect of biodiversity to focus on: species diversity, genetic diversity, functional diversity? The specific objects to be protected are not necessarily the same.

The third aspect is relating to the *effects of conservation* again (as with protected areas). Ecosystem services are certainly beneficial for some people but may also generate costs and burdens for others. Thus, e.g., the production of energy crops imported by highly industrialised countries (provisioning ecosystem services), whose people and companies benefit from them, threatens the livelihood of people elsewhere, e.g., through destruction of traditional landscapes and the heavy use of pesticides. Therefore, questions of justice arise. Often different groups of people are affected either beneficially or negatively by specific management decisions, including conservation management. There can also be conflicts between the use of ecosystem services and the protection of biodiversity or other human activities.

The distribution of benefits and costs associated with the provision of ecosystem services, but also with other conservation measures, is important across all scales, both spatial (e.g. global gain – local loss – or vice versa) and temporal (use by present generations or options for future generations).

Conservation, economic structures, and radical critiques

The potential implications of the ecosystem services concept and other conservation concepts on questions of power and justice lead the gaze towards more fundamental critiques that have been raised about the current theory and practice of conservation.

In the foreword of the book derived from the 1986 "Forum on BioDiversity" (see earlier), its editor Edward O. Wilson celebrated the "new alliance between scientific, governmental, and commercial forces", which he "expected to reshape the international conservation movement for decades to come" (Wilson 1988, p. vi).

This in fact was true, because already the CBD shifted the focus beyond "mere" conservation to include not only the links towards human well-being (in the broadest sense) but also to the links between conservation and the economy – in order to mainstream biodiversity protection.

The relation towards the economy, and partnering also with "big business", i.e. large companies, for the (presumed) sake of conservation has been a major point of dissent in the debate on the "new conservation" (see Section 2.8). The most radical critique of economic language and thought in conservation, and of conservation approaches embracing the traditional capitalist structures and logic, usually comes from political ecologists (e.g. Büscher and Fletcher 2020). But also scholars who have been deeply involved in research on ecosystem services (e.g. Muaradian and Gómez-Baggethun 2021) have called for radical changes with respect to the problems of the capitalist logic that most conservation still builds on. It is worth looking at these arguments, the more as they show that the perceived dichotomy between "old" and "new" conservation is not as clear-cut as it seems and is also not the only major distinction between current conservation trends.

Büscher and Fletcher (2020) describe, as a first approximation, two major axes along which conservation approaches can be classified: one reaching from capitalist to "beyond-capitalist" positions, the other from positions "steeped in nature-people dichotomies" (p. 7) towards those that go beyond such dichotomies.

Interestingly, they argue that not only the "new conservation" is building on the logic of capitalism but also what they call "mainstream conservation". The latter for them includes many classical ideas focusing on protected areas as well as more recent participatory approaches, including community-based conservation.

They see mainstream conservation as "steeped in nature-people dichotomies" while "new conservation" is seen as *not* embracing this dichotomy. As a third type of conservation they see what they call "neoprotectionism", a more recent trend arguing for the necessity of the separation between humans and nature to save the Earth's life-support systems (and biodiversity), such as the "Half Earth" movement. Büscher and Fletcher perceive of many, if not most neoprotectionists as critical to contemporary capitalism. The authors' own alternative, which they call "convivial conservation" and which they develop in the last part of the book, is seen as both critical to capitalism as well as overcoming the nature–culture dichotomy. They argue:

> To put it bluntly: without directly addressing capitalism *and* its many engrained dichotomies and contradictions, we cannot tackle the conservation challenges before us.
>
> *(Büscher and Fletcher 2020, p. 9)*

As the subject of human–nature dichotomies was discussed at some length already earlier, I will here largely focus on the issue of the capitalist logic.

The points made by Büscher and Fletcher partly match with those discussed by Agrawal (1997), i.e. they deplore a problematic focus of conservation ("mainstream" and "new conservation") on progress and development, related to the idea of unfettered growth and consumption, especially by people in the rich countries. This again is connected with a highly utilitarian perspective on nature, expressed earlier in the narrative of "Safeguarding necessary and useful nature". Some conservationists, who clearly reject human exploitation of nature and call for more protected areas, are at least highly committed to the idea of technological progress. E.O. Wilson (2016), postulating to reserve "Half Earth" for the protection of non-human species, is at least critical on the high per capita consumption of humans and on population growth, but he does not mention the disproportionate effect of global elites and rich countries on resource use and related destruction of habitats. He instead (chapter 20 in Wilson 2016) postulates a transformation of the economy from "extensive economic growth" to "intensive economic growth", thus not questioning the general idea of growth. More than that, he embraces a kind of 21st century technocratic optimism, promoting a "linkage of biology, nanotechnology, and robotics" (p. 195) and "entrepreneurial innovation" (ibid.), with synthetic biology being part of the solution to conservation problems. What comes up also as an argument again, with Wilson and others, is the idea of conservation as an issue of civilisation,[2] by which it may be inferred without too much distortion, "Western" civilisation is meant.

A critique of the elements of Western capitalist logic (progress, growth, utilitarianism) and sometimes of the capitalist economic system in total is also explicitly expressed by non-Western societies, e.g., in the context of "buen vivir" (see Section 4.4).

The critique on the capitalist logic as expressed by Büscher, Fletcher, and others has in itself been criticised and relativised by some authors, claiming that the former miss out some important points (e.g. Kopnina 2016, Cafaro et al. 2017). Such points are seen, e.g., in the problem of population growth, which some conservationists consider as more important than the high per capita consumption of rich countries and global elites, or in the allegation that justice arguments were mainly reduced to those of justice between humans and did not account enough (if at all) for justice between humans and non-human beings.

A rejection of capitalist "solutions" to conservation on the grounds sketched above does, however, not mean to refrain from instrumental arguments to nature conservation. It only demands that social inequality and the unfettered exploitation of "natural resources", perpetuated by the current economic mainstream, must be acknowledged as one, if not the main reason for the environmental crisis (Matulis & Moyer 2017). A critique of the prevailing social and economic conditions, including the still dominant focus on growth and progress, and unequal power relations, is therefore necessary both for the sake of humans and for that of non-human beings.

Is there a common "we" in conservation? No and yes

Arguments on the role of power and justice lead back to the question as to whether there is a common "we" in conservation. Political scientists and sociologists have often emphasised that there is no such a simple "we" in terms of "humanity" when talking about conservation. I briefly touched upon this already in Section 2.6 but it is in the context of the implementation of conservation concepts that this issue becomes crucial. Also the considerations on power and justice demonstrate that in many cases it is problematic, in social and moral terms, to speak of such a common "we". Not recognising that there are often multiple actors (or actants in the language of narrative analysis introduced in Section 3.2), which are in different ways acting and/or being affected, can cause conflicts, both within the conservation community as with those affected by conservation measures. The different groups to be considered may be those of different income, different education, different gender, or different culture. Of course, the lines among them are not sharp and roles may change depending on the specific context – and trough time. Simple dichotomies such as Western and non-Western and scientific and indigenous, are often not helpful. Even within specific groups, perceived as homogenous by an outsider, there can be significant power relations.

The critique of political scientists and sociologists that the societal differences, especially differences in power and standing, are neglected too often in conservation is justified in many cases and is a point that requires explicit consideration. But in fact it did not go unheeded within the conservation community. For instance, the rivet popper story, brought forward by Ehrlich and Ehrlich (1981) as a metaphor to illustrate the consequences of species extinction (see Section 3.2), implicitly sees *humanity as a whole* as the passengers of the vulnerable airplane (= planet Earth, or its biosphere). In an even earlier text, however, Ehrlich extended the metaphor, stating:

> It is as if the poor of the world own the wings of our aeroplane and one of the things they must do is pry the rivets from the wings one by one and sell them. It would clearly be worth the while of the first class passengers to share the wealth with the owners of the wings so that the poor would not have to sell the wings to survive.
>
> *(Ehrlich 1980, p. 343)*

Agrawal, who cites this passage in his paper from 1997, however, criticises these and similar statements as putting the blame of environmental destruction on the poor, neglecting the reasons of poverty. Now the "West" and its superior science and economy (which to a considerable degree caused environmental problems) would give support to the developing countries for sustaining the Earth's life support systems. For Agrawal this is still a continuation of colonial attitudes. So clearly a common "we" in conservation is largely an illusion, also when it comes to different ideas of nature and good human–nature relationships.

An interesting caveat should, however, be mentioned here. Without neglecting the important distinctions and differences between human actants, some authors have recently pointed out that in the face of global anthropogenic impacts (the "Anthropocene") and especially in that of climate change, humans are now also threatened as a *species.* Normally humans do not really consider themselves as a species (unless they are biologists…). As Dipesh Chakrabarty (2009, p. 220) writes:

> We humans never experience ourselves as a species. We can only intellectually comprehend or infer the existence of the human species but never experience it as such. There could be no phenomenology of us as a species. Even if we were to emotionally identify with a word like *mankind*, we would not know what being a species is, for, in species history, humans are only an instance of the concept species as indeed would be any other life form. But one never experiences being a concept.

But in becoming a "geological agent" and causing even climate to change, a self-perception as a species becomes something necessary in his eyes:

> Climate change, refracted through global capital, will no doubt accentuate the logic of inequality that runs through the rule of capital; some people will no doubt gain temporarily at the expense of others. But the whole crisis cannot be reduced to a story of capitalism. Unlike in the crises of capitalism, there are no lifeboats here for the rich and the privileged.
>
> *(ibid., p. 221)*

And he concludes:

> Species may indeed be the name of a placeholder for an emergent, new universal history of humans that flashes up in the moment of the danger that is climate change. But we can never understand this universal.
>
> *(ibid., p. 221f)*

This idea produces an interesting tension which one should keep in mind when thinking about conservation and human–nature relationships. The question of who exactly are the actants in specific conservation narratives remains important though; considering it more clearly may reshape the narratives, especially as global change proceeds.

A point we should take from the discussion on the "we" in conservation, by explicating the different actants, is that it helps to better understand conservation conflicts. Conflicts are not always a bad thing but, taken seriously, may start explicit discourses where decisions would otherwise be simply based on – and perpetuate – existing power relations (Redpath et al. 2015a, Matulis and Moyer 2017).

5.3 Arriving at good relationships with nature: criteria and requirements for selecting appropriate conservation concepts in specific settings

Which further lessons can be taken from the previous chapters for choosing and applying conservation concepts in a reflexive way, leading the path towards good relationships with nature? Building on the results of the previous chapters, some criteria for choosing appropriate conservation concepts emerge. Formulated as questions, these are as follows:

- What are the ultimate target objects?
- What are the biophysical boundary conditions?
- What is the specific social setting (stakeholders, values, and power relations)?
- What is/are the desired role(s) of humans with respect to nature?

On top of that is another question, which up to now has been dealt with only in passing, which is:

- What are the purposes of using conservation concepts?

In addition, some pragmatic criteria (such as the required precision of definition) have to be considered. These are largely dependent on the last question (purposes of using conservation concepts).

The criteria cannot be presented as a simple checklist with various options for each criterion and a list of concepts that fulfil them. Especially the biophysical and social conditions prevailing are not easily classified. For some criteria (target objects, desired roles of humans) the tables in Chapter 3 can provide some orientation. Also, these questions should not be dealt with in isolation, one by one. Instead of presenting such a checklist and explaining the criteria one by one, I will therefore describe a set of requirements that should guide the selection of conservation concepts under the perspective of good human–nature relationships. Together with what has been elaborated before, and partly reiterating them, these requirements should fill the above criteria with substance and *connect them*. Section 5.4 will then condense the criteria elaborated in this chapter into a brief suggestion for a procedure – a sequence of steps – for selecting appropriate conservation concepts in specific settings.

Different purposes using conservation concepts

The specific purposes for which conservation concepts are used have a critical influence on the selection of concepts, especially with respect to some methodological requirements (pragmatic criteria). The main purposes for which conservation concepts are used are communication and justification of conservation goals, assessment of the state of nature and of conservation values, and their use for guiding the implementation of conservation goals. These purposes decide, e.g., about the required degree of precision needed in the definition of the concept. If the ecosystem services concept is, e.g., just used for didactic purposes, communicating the value of nature in general, it requires only a broad definition; but it needs a very specific one if used, e.g., for quantifying the services of a specific place. Other pragmatic criteria related to the purpose of using a concept are, e.g., if the application is focused on specific areas or not, or the degree of generality and transferability between different places which the concept allows for.

In the following, five more content-related requirements are described, connecting the criteria formulated above.

The relational imperative: embracing relations as a crucial category for conservation

Why put so much emphasis on the relational in the selection and application of conservation concepts? Of course this book is dealing with human–nature relationships. And of course even the most destructive ways of humans dealing with nature point to relations, such as those of the heedless exploitation of ecosystems as merely "natural resources". This, however, is not what is meant here by the "imperative of the relational". As described already in Section 3.1, there has been a recent debate on "relational values". In this debate (and earlier predecessors) an important distinction is made between those relations where the *objects* of the relation are in the focus, e.g. a tree as a resource for improving human well-being via timber, and those where the *relation itself* is the focus, is what is valued (see

Himes and Muraca 2018 for problems of terminology). Relational values in this sense are then contrasted with use values (or instrumental values) on the one side and intrinsic values on the other. Such relational approaches are not limited to those between human and non-human nature but also refer to those between humans.

Figure 5.4 shows how the difference between the value types plays out. Instrumental values as well as intrinsic values place a distance between humans and nature, the former using nature as an object, the latter distancing humans from nature for the sake of nature. In contrast, relational values bring about a much closer two-way relation between nature and humans– both the individual as well as groups of humans, – and indirectly mediate relations between humans. This reinforces the point that human–human and human–nature relations are intimately

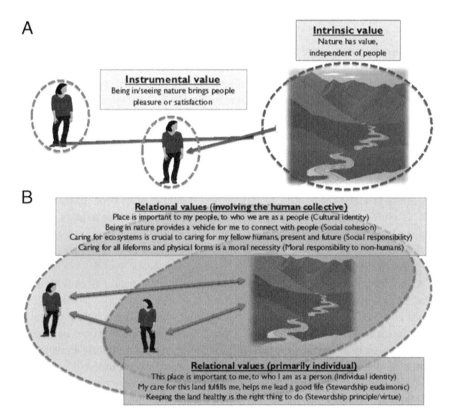

FIGURE 5.4 The difference between the instrumental and intrinsic value framings that dominate environmental literatures and relational values. See text for further explanation.

Source: Figure reprinted from Chan, K.M.A., Balvanera, P., Benessaiah, K. et al. (2016). Why protect nature? Rethinking values and the environment. *Proceedings of the National Academy of Sciences of the United States of America*, 113, 1462–1465, with permission of PNAS.

connected. One and the same person can have all three types of relations to nature, those referring to utility, those valuing nature as distinct and independent from oneself, and those where the relation as such is decisive, as captured in the idea of relational values. Also, using nature in an instrumental way is not morally bad, but it is necessary for any human life. The same can be said for relations with other humans. We *also* use humans in an instrumental way, but, as Kant said in the "humanity formula" of his categorical imperative: "Act so to treat humanity, whether in your own person or in that of any other, at any time *also* as an end, and never *merely* as a means"[3] (Kant, Metaphysics of Virtue 429; my emphasis).

So an instrumental relation does not preclude other, non-instrumental ones. People, value and love their spouses not only because they think them to be of instrumental value only (e.g. preparing breakfast, earning money for living), neither only because the other person is seen as having intrinsic value. What counts is the relation as such, and that is what relational values, in the meaning used here, refer to.

Relational values come much closer to the intuition and experience of most people, are much closer to everyday experiences than the more abstract intrinsic values. Promoting human–nature relationships and conservation concepts which embrace this type of value thus may also be of strategical importance for conservation.

I want to emphasise here that relations do not always have to be expressed in the language of values; my impression is even that the value perspective in conservation is sometimes already oversold. Relations may also be expressed in a less abstract way through behaviour/practice and notions of personal identity as well as group identity. And in this respect, some conservation approaches as well as philosophies and practices converge with the now increasingly popular relational approach in conservation (see, e.g. a special issue in Current Opinion in Environmental Sustainability in 2018,[4] but also Berghöfer et al. 2022, Hoelle et al. 2022, Pascual et al. 2022).

As one point of convergence, relational approaches to nature align with many ideas in indigenous and traditional thinking (e.g. Peterson 2001, Escobar 2016, Gould et al. 2019; see also Sections 4.3 and 4.4.). Also Tim Ingold's notion of "dwelling" can be seen as a relational approach, emphasising practices of humans and human embeddedness in nature (Ingold 2000, see also Obrador-Pons 2007). In many indigenous societies "nature" is seen not just as resource but something in which non-human beings (also spiritual ones) interact in complex relations with humans. In recent Western philosophy, feministic ethics and care ethics also emphasise relational approaches to nature (e.g. Warren 2015, Jax et al. 2018). Likewise, as mentioned earlier (Section 4.4), philosopher Martha Nussbaum has emphasised relations as fundamental for a good and well-lived human life, including relations with non-human nature. Of course this idea is present also in psychological literature on the issue of human–nature relationships (Wiggins et al. 2012).

I think that the whole concept of relational approaches to (or perspectives on) nature has a huge potential for conservation, which still has not been used fully and which needs further development. For me personally, pursuing such an approach is not just a strategical issue but a real moral *imperative*, both with regard to humans as to nature, connecting both. I think we need to move forward with nature, or at least with non-human living beings, in the way that the Jewish philosopher Martin Buber suggested: from an "it" to a "thou".[5] For Buber, there are two different modes of relating to the world, with the relation "I-it" as considering something as external to myself, as a mere object, while "I-thou" (or "I-you") constitutes a relation to a counterpart: "The world as experience belongs to the basic word I-It. The basic word I-You establishes the world of relation" (Buber 1923/1970, p. 48f).[6] This links to some indigenous ideas of respect for nature but it is in no way dependent on mystical or religious assumptions. It means to add a relation to non-human nature that goes far beyond dealing with it as an object, as an object *only*. Buber does not reject other approaches to nature, the need to perceive of other living beings *also* as an "it" (see Buber's description of perceiving a tree, Buber 1923/1970, p. 49f). Nor do the Cree Indians (Section 4.3) and other indigenous peoples. Nor should a relational approach for conservation, as discussed here and in the literature, aim at replacing instrumental uses of nature or its perception as intrinsically valuable. Instead it supplements it in a crucial and necessary way.

What does that mean for conservation narratives and for using conservation concepts? How does or can a relational approach surface in narratives of human–nature relationships as given in conservation? Before I try to answer this question, let me remind you about the difference between the narratives of human–nature relationship described in Section 3.3 and the concepts in the conceptual clusters in Section 3.4. Although I aligned them, referring to each of the concepts that designate the clusters also as paradigmatic for one of the narratives, they are not the same. The narratives are meant as kind of "ideal types", partly even caricatures, of the ideas of human–nature relationships implicit in conservation concepts – in order to sharpen the contours of these relationships. The narratives are nevertheless a common way in which the "paradigmatic concepts" (e.g. biodiversity, ecosystem services, wilderness) are perceived by people, even though each of the concepts is much more complex in reality, has undergone many changes, and is subject to many different interpretations and often branching into "sub-concepts" (such as species diversity and functional diversity for biodiversity).

For the narratives introduced in Section 3.3, the extreme positions are those of "Nature apart from humans" and "Safeguarding necessary and useful nature" (both in their "hard-core" versions). They leave no room for this kind of relational approach and they, in a way, stand for the assumed hiatus between an ecocentric (or physiocentric) ethic on the one side and an anthropocentric ethic on the other (as depicted in Figure 5.4). But in between these extremes there is a wide spectrum of other human–nature relationships. All the other narratives of human–nature relationships (as discussed in Section 3.3) in principle leave room for relational approaches. The strongest candidates here are "Caring for other living beings…" and "Being at home in nature".

Much hinges on different ideas of what humans are and which role they have and should have with respect to non-human nature, as expressed in the narratives. It thus also depends on what is understood by (proper) nature and *how* that nature is (see below). Are humans (or "man" as Wilson put it) in principle "destroyer[s] of the living world", "arrogant, reckless" (Wilson 2016, p. 1)? Is (non-human) nature already "spoiled", less valuable, once humans interact with it? Or are humans stewards, co-inhabitants of nature, or may, in part, even improve on it, create new value within nature?

To conclude, I suggest that in pondering on good human relationships with nature and in selecting appropriate conservation concepts, we should always include relational approaches in the sense discussed here. Approaches such as ecosystem services, not the least promoted for strategic reasons, are important, but they must be supplemented by other approaches, foregrounding relational values (and as I think, "respect for nature"). In spite of all honest efforts to avoid this, ecosystem services and similar concepts from that cluster will always be associated mainly with the extreme "Safeguarding necessary and useful nature" narrative and the current capitalist economic system, based on growth and progress (see Muradian and Gomez-Baggethun 2021). Beyond that, human–nature relationships, as implied in relational approaches, must not be restricted to relations between humans and non-human nature but also extend to relations among humans. Doing this is necessary and productive for all actants involved (including non-human ones) but raises questions of balancing different needs and demands.

Balancing the needs and demands of (different groups of) humans and non-human species

The obvious (and not at all novel) demand for the selection of appropriate conservation concepts in specific settings is to scrutinise for potential conflicts that may arise when implementing them.

There is a broad literature on tackling conservation conflicts (e.g. Wittmer et al. 2006, Redpath et al. 2013, 2015b). I will therefore not go into every detail here but only highlight some points of special importance in the current context.

Conservation conflicts are in the first place not conflicts of humans with nature but conflicts between different groups of humans with different ideas of good human–nature relationships and human well-being. As said before, in the question of what makes "us" human, relations between humans and relations with non-human nature are closely intertwined. Conservation conflicts between humans affect not only the needs and demands of humans but also those of non-human nature. This brings up questions of power and justice, as discussed in Section 5.2. It is important to emphasise once more that conservationists are not only "stakeholders" in conservation conflicts but that often *different groups of conservationists and conservation biologists* disagree about what are the proper conservation concepts to be followed.

Conservation conflicts are not only about material or even economic interests about distributional justice. Of course such issues play an important role, both between different groups of people affected by conservation as with regard to non-human organisms. But in addition, and especially between different groups in the conservation community, it is about values, about different ideas of nature, different ideas the role of humans in nature – summed up as good human–nature relationships.

When considering a particular conservation concept in a given setting, it should, as a first thing, not simply be taken for granted as the only applicable one. Instead, there should be openness for discussing and responding to potential problems of the respective approach in terms of environmental and ecological justice, and, if necessary, potential alternatives should be considered, given the diversity of conservation concepts and attitudes.

What is necessary is a critical screening of the actants involved, whose needs and demands might be furthered but also possibly be compromised by the application of the approach (e.g. when establishing a protected area). For humans, these needs and demands include material ones as well as non-material ones related to ideas of a good human life, including good human–nature relationships. For non-human entities needs and demands obviously relate to their physiological, cognitive, and ecological requirements. When accounting for the needs of non-human entities, it is, however, often unclear, which features of nature, which *objects* are to be considered. This varies with the specific conservation concept and type of human–nature relationship embraced (see tables in Chapter 3). Is it individual organisms, is it species (as represented by individual organisms and their populations), is it ecosystems, or even "nature as a whole"?

It is highly controversial whether entities such as ecosystems are directly morally relevant or if needs and interests can be ascribed to them. Nevertheless, there have been attempts to provide at least *legal* (even though not necessarily moral) rights and standing to such supra-organismal entities. While still an exception, there are increasing efforts in this direction, often connected with establishing juridical personhood (also called environmental personhood) for such entities (see, e.g. Gordon 2018, Miller 2019). Famous examples are the Whanganui River in New Zealand and the Gangotri and Yamunotri Glaciers in India, which have been declared as legal subjects in the jurisdiction of the respective countries. Even though I personally doubt that ecosystems or landscapes can have moral rights, the attribution of legal rights might be one interesting way to "give a voice to nature".

The ideal will be that the conservation concept selected will promote integrative human–nature relationships that can satisfy all the different needs, human as well as non-human ones. However, even with careful consideration of all options, the implementation of conservation approaches does not always produce win-win solutions but must face the fact that there may be winners and losers, both among humans as among non-human entities (Chan et al. 2007, McShane et al. 2011, Muradian et al. 2013). These winners and losers must be identified. Potential gains and losses must be made explicit and solutions should be sought to compensate those who lose, in the fairest and most acceptable way. Some conflicts, however, will not be resolved.

Sometimes different ideas about proper human–nature relationship may be combined and realised jointly or at least in the form of a good compromise. As conservation is mostly local or regional in its implementation, it will, however, in many cases be possible to account for different human–nature relationships by spatial separation, meaning that, e.g., wilderness areas and cultural landscapes can coexist on a larger scale. UNESCO's concept of the Biosphere Reserve (developed in the 1960s, implemented first in the 1970s; Bridgewater 2016) is an excellent example here. The zoning scheme of Biosphere reserves allows to account for different types of human–nature relationships in different parts of the designated area. The *core area* must be a strongly protected area such as a national park and follows largely the narrative of "Humans apart from nature", combined with "Caring for living beings…". The other zones do not require legal protection. The so-called "*buffer zone*" is described by UNESCO as "used for activities compatible with sound ecological practices that can reinforce scientific research, monitoring, training and education" (https://en.unesco.org/biosphere/about). Finally, and perhaps most interestingly, the so-called "*transition area*" is characterised as an area "where communities foster socio-culturally and ecologically sustainable economic and human activities" (ibid.) and thus can lend itself, e.g., for the relation of "Being at home in nature" or "Steering nature in a responsible way". Concepts like the Biosphere reserve have great potential to "harmonise" different ideas of nature and good human–nature relationships. An example may be the Biosphere Reserve "Berchtesgadener Land" in the very south of Germany (see https://brbgl.de/). The core zone is here constituted by the national park Berchtesgaden with strict protection (IUCN category II), while the buffer zone is devoted to maintaining a low-impact cultural landscape. The transition zone (here called "development zone", *Entwicklungszone*) grades into the normal rural (cultural) landscape and small urban areas, where efforts to develop sustainable ways for living and economy are actively promoted.

Accounting for biophysical and social constraints at relevant scales

Many international conventions and institutions focusing on nature conservation show how much conservation is a global issue meanwhile. Nevertheless, the implementation of conservation measures is mainly taking place on local or regional scales. Also, the effects of biodiversity loss on ecosystems are mostly local, at the scale of specific ecosystems, not global. It is thus the conditions (and their history) that prevail in a specific location that constrain the choice of conservation concepts, or at least should do so.

BOX 5.3 PROTECTED AREAS: AREA-BASED CONSERVATION IS MORE THAN FORTRESS CONSERVATION

Protected areas have been the "dominant big idea" of 20th-century conservation (Adams 2004, p. 4). It has, however – and sometimes rightly so – often been said that protected areas stand for "fortress conservation" (see Section 5.2) excluding people from areas and from nature, and even depriving people of their traditional areas and ways of life. But even though national parks and wilderness areas, wherein the role of humans is minimised, attract most attention from the public, protected areas do not necessarily imply that humans are excluded. There is a broad spectrum of protected area types, in which partly even specific human–nature interactions are to be protected. In any case, however, the designation of some area as protected restricts at least *certain kinds* of interactions/land uses. The diversity of protected areas types is reflected in the IUCN protected areas categories (Dudley 2013), which reach from "Strict nature reserve" (category Ia) and "Wilderness area" (Ib) over "National park" (II) to "Protected landscape/seascape" (V) and "Protected area with sustainable use of natural resources" (VI). Especially the latter two categories explicitly include the roles of humans (historically and present) as co-creating these areas as valuable for nature conservation. Thus, the definition of Category V reads:

> A protected area where the interaction of people and nature over time has produced an area of distinct character with significant ecological, biological, cultural and scenic value: and where safeguarding the integrity of this interaction is vital to protecting and sustaining the area and its associated nature conservation and other values.
>
> (Dudley 2013, p. 20)

Even broader, beyond IUCN's categories, is the concept of UNESCO's Biosphere Reserves, with its zoning allowing for different types of human–nature relationships (see earlier).

The target of the of CBD's recent "Kunming-Montreal Global biodiversity framework" postulating to protect 30% of the Earth bei 2030,[7] must thus be read carefully. It does not postulate 30% of *human-free* zones but covers a much broader set of protected area concepts, also those including humans and their activities.

As to whether the designation of a specific protected area is problematic for local people depends, inter alia, on how much, or if at all, these people have been heard and involved in the process of selecting and implementing a particular type of protected area and its management – which, however, is still often not the case.

This local perspective is also important for the question which I found the most difficult to answer for the narratives described in Section 3.3, namely *how nature is*. Is nature, e.g., robust or fragile? The answer depends partly on what we specifically mean by nature or on which aspects of nature we are focusing. Understood as the whole of the living biosphere, of course, "nature" will continue to exist, even if only some bacteria and algae will survive human onslaught. In this way, it is robust, but it may not be the nature we cherish, even not be a nature conducive to human life, not to mention a good and comfortable life. But the judgement as to whether nature is robust or fragile (or which other properties it has) depends on the specific part of nature we look at and our idea what an "intact" (or functioning) nature is (see Section 3.4). It is thus something that must be tackled in specific situations and places, without too general *a priori* assumptions of robustness, fragility, and resilience. Sometimes these "places" can amount to large and even spatially discontinuous areas, e.g., when it comes to saving large carnivores or migratory species.

It is the very role of science, especially ecology, to describe what kind of ecological setting is given or possible in a location (polar bears will hardly do well in the tropics, desert plants not in boreal lowlands), which components and processes are necessary for particular ecosystems or for the thriving of specific species. It is a major challenge for restoration efforts (and the implementation of specific restoration concepts) to clarify to which state a "degraded" area *can* be developed; can we really go back to the "original" state of nature on a site? But also all other conservation concepts face the question of what kind of nature is feasible and what it takes to maintain them in terms of management. Even if conservation or restoration is feasible in a purely biophysical and technical manner, other constraints may apply on behalf of the economic and/or social costs of it – something to consider in the face of (always) limited funds and work power.

On the knowledge side of ecology and conservation biology uncertainties of several kinds (Haila et al. 2014) will remain, as is normal in science, also idiosyncrasies exist, which make every site and situation to some degree special. In particular on the local level, traditional and indigenous knowledge, if existing, can be of high importance for conservation, often even be superior; conservation is not about scientific "truth" but about understanding and practice.

Heeding the local situation is necessary for another reason that has been touched upon already: in many conservation contexts *place* is not simply *space*, but something where people are attached to in terms of non-material, often emotional relations and identity (personal as well as collective). This is prominent in many local and indigenous perspectives. But also different conservation concepts and narratives have different takes on this. Thus for the relation of "Being at home in nature" and for most concepts of cultural landscapes, the "sense of place", the attachment to particular places as a result of human interaction with nature, is essential. Also for "Caring for other living beings..." and for many understandings of "biodiversity", a relation to the "typical" and thus place-specific is important. Much less importance, sometimes even none, does place have for the "Safeguarding necessary and useful nature" relation and many interpretations of the ecosystem services concept, which have much more a global perspective or at least one not bound to specific places (see Jax and Heink 2015). Likewise, place is, at least at first, not as important in the concept of novel ecosystems, but it may gain importance if such an ecosystem becomes older and more appreciated.

Emphasising the local level in effect also means to involve local people in decision processes, also in the selection of the most appropriate conservation concepts.

Means and ends: conservation concepts should not be strategic only

Conservation is political, not just academic or private. When conservation scientists and conservationists address politics, particular societal groups, or the public at large, strategic considerations are often one criterion in the selection of arguments and concepts. It is prudent and necessary to account for a broader spectrum of audiences beyond those who are already convinced about the need for conservation, to adapt one's language and arguments accordingly. However, as discussed in Section 5.1, there is a danger in focusing almost exclusively on one argument for strategic reasons. I discussed that for the concepts belonging to the cluster "ecosystem services" (Section 3.4), building to a considerable degree on the narrative of "Safeguarding necessary and useful nature", a point which has been criticised repeatedly with respect to its closeness to Western capitalist thinking. In a survey about the attitudes of more than 9000 conservationists from around the world Chris Sandbrook and colleagues (2019), however, found that 97% of them agreed to the statement that "Maintaining biological diversity should be a goal of conservation" (whatever that means in detail), with only 24% being supportive of the statement that "The best way for conservation to contribute to human wellbeing is by promoting economic growth". There is a danger that arguments based largely on strategic considerations become problematic once they become dominant. They may be oversold, do not reflect the whole spectrum of conservation concepts and human–nature relationship they refer to; those other human–nature relationship may become neglected in public discourse. In addition, controversies within the conservation community ensue. To still use the strategic dimensions of conservation concepts without drawbacks it is necessary to be clearly aware of the differences between the means and the ends of conservation efforts. Strategic uses focus on the means (e.g. convincing people with a particular argument, using a wilderness area to protect certain species, applying restoration to enhance ecosystem services or biodiversity), but the ends – as reflected both in the ultimate conservation objects as in a specific desired human–nature relationships (or a set of the same) – should be clear and eventually be guiding conservation.

There will be no silver bullet, in the form of a concept that integrates all types of human–nature relationships and potential conservation targets (objects). A better way is to acknowledge the plurality of conservation concepts and human–nature relationships and make use of them in a prudent way. On the strategic level alone it is important to reach as many people as possible with conservation arguments. More than that, acknowledging plural relationships and values associated to them is a matter of respect to those people involved in and affected by conservation, and, even more, of justice. It is also a matter of transparency and truthfulness in the public arena (Jax and Heink 2015). The plurality of concepts may even be extended by carefully using other knowledge types currently being subordinate to scientific and economic concepts, such as indigenous and local knowledge (ILK), a difficult task, as discussed earlier, but a necessary one if conservation science is not to remain a purely Western endeavour. Many such efforts are underway, exemplified by the processes within IPBES, e.g. the creation of a Task Force on ILK.

It is also the narratives on human–nature relationships themselves – beyond the specific conservation concepts – which are often used in a strategic manner, where the most "effective" narratives for conservation are sought. But the same problem as discussed pertains also here.

Multiple human–nature relationships should be accounted for

One cannot simply *choose* a human–nature relationship, as one can chose a conservation concept or a way of living. We either already *have* a particular human–nature relationship (however conscious) or we can *strive* for it. Likewise, it is difficult to simply *invent* completely new narratives of human–nature relationships or transfer such narratives from other cultures; resonance to something already existing in our cultures and ways of life is at least useful, if not necessary (Peterson 2001, p. 230f), even if these are shared only by minorities. What we can do is to communicate narratives foreign to us and take them for reflecting our own ideas and experiences. There are also some material constraints: we (as Westerners) can, e.g., not simply go back to the intimate direct experiences of hunter and gatherer societies (Strang 2005), but we live in contexts where our everyday objects (including food) are brought to us largely in an indirect, mediated way. Ideas of desired human–nature relationships, however, can and should guide us when we search for appropriate conservation concepts and ways to implement them.

These *desired* human–nature relationships are described by the last stages of the narrative scheme developed in Section 3.2, namely that of the desired future, including the desired role of humans with respect to non-human nature (see also Table 3.1).

What does it mean to account for multiple human–nature relationships? I assume that almost all conservationists subscribe not only to one of the human–nature relationships described by the 6½ narratives in Section 3.3 but to several of these.

Nobody will deny basic human needs and their legitimacy and almost all people will argue for the sustained existence of humanity,[8] with a decent level of human well-being. So it can be taken for granted that almost all conservationists, however implicitly, embrace the narrative of a "Safeguarding necessary and useful nature".

But the question is, which expression of this narrative: is it mainly about survival, about a good human life in the Aristotelian sense (eudaimonia; Section 3.1), or about economic prosperity? This is where conflicts arise.

Likewise, the wilderness concept can be turned from an end into a means. In that way it may not necessarily embrace a radical "Humans apart from nature" narrative, but one in which having areas where humans do not intervene is an expression of respect towards other living beings (narrative "Caring about other living beings…"). As far as people are included in such areas as dwellers, there is a gradient towards the "Being at home narrative". The latter, e.g., comes much closer to the perception of many indigenous people, who – in contrast to many conservationists – do not perceive of their environment as "wilderness" but more as a kind of "cultural landscape". Cultural landscapes, then, also can account for many of the values embraced by the narrative "Caring about other living beings".

The biggest problems of disunity among conservationists arise when *purity* for one of the conservation narratives is sought, such as the "pure" wilderness, the "pure" ecosystem services approach, the "pure" cultural landscape, the "pure" *native* biodiversity. Often such purity is in fact not really proposed by those adhering to some concept but by their opponents, such as when fighting the "pure novel ecosystems" idea, assumed to be meant as a substitute for classical conservation approaches. Section 2.8 already illustrated how strawmen were created by both supporters as opponents of the "new conservation" in depicting the respective other side, both sides striving for "true" conservation.

Even if they do not strive for such purity, conservationists should be aware of the messages, the implicit narratives that are transported when they argue for specific concepts. Even if these connotations are in fact contrary to their much more differentiated understanding of concepts, it might alienate other conservationists and other people from their goals. Sober explanation and discussion is necessary.

So conservationists can and mostly will embrace multiple human–nature relationships. It will nevertheless always remain a challenge how to account for such multiple human–nature relationships and balance them in specific conservation tasks and settings. Which of them should take priority? Or can they be balanced in a way that satisfies several of them in an equal manner? In any case, acknowledging the validity of multiple conservation concepts, to be applied in specific situations and areas, is not relativism and selling out on the ideals of a favoured conservation concept, it is accounting for the complexity of nature, human societies, and their mutual relationships.

With all carefulness – as he has been interpreted by so many scholars and in so many different ways – Aldo Leopold appears to be a very positive example of how one person can bring together different narratives on human–nature relationships in his life and work, and this in a very reflexive way. Leopold was both a scientist, an experienced practitioner of conservation and natural resource management, and a thinker. Having been educated and working as a wildlife manager and forester, the "useful and necessary" character of nature was obvious to him. He was nevertheless also one of the founders of the Wilderness Society in the USA, seeing much value in wilderness, where humans were only guests. In addition, Leopold's idea of the "biotic community", or "land community", a major concept in his Land Ethic, was clearly valuing each member of that community (Millstein 2018), and explicitly included humans in that concept of the land community. His ideas in this respect were somewhere between perceiving of the land (nature) "as home" and as the duty of steering nature responsibly (ecosystem management in today's parlance) (Meine 2020). In addition, Leopold was a pioneer of ecological restoration, striving for healing "a world of wounds", contributing to "making nature whole again" (see Section 3.4). My impression is that Leopold was in no way puristic or dogmatic, given the variety of his relationships and practical approaches to nature. He was aware that different tracts of the land would be handled differently, also from a perspective of conservation.

5.4 A procedural suggestion to clarify conservation goals and to tackle conservation conflicts

This section condenses the criteria and requirements elaborated in Section 5.3 into a brief suggestion for a procedure – a sequence of steps – for selecting appropriate conservation concepts in specific settings. This selection, based on ideas of good human–nature relationships, should serve not the least as a means to clarify conservation goals and approaches and to better tackle conservation conflicts.

Most of the steps presented in the following are not completely new compared to what has been discussed elsewhere with respect to analysing and dealing with conservation conflicts (or avoiding them).[9] It is however, not meant to be used in isolation but must be read in the context of what has been elaborated in Sections 5.1–5.3. It is there that the specific rationale for the different steps and details of things to consider have been expounded. In the face of diverging ideas between conservationists, the procedure should provide a basis for a more rational discussion towards selecting appropriate conservation concepts.

1 **Purpose.** First, the specific purpose for using specific conservation concepts should be clarified.

Are the concepts (e.g.) meant as a communicative device, a justification, a lever, an assessment tool, or a practical guidance for conservation? Specific concepts may be adequate for one or some of these uses, but problematic for others.

Scrutinise pragmatic criteria with respect to this purpose (as described at the beginning of Section 5.3).

2 **Stakeholders and other actors.** Identify stakeholders (in the widest sense) for the issue at hand *and include them in all further steps*. Conservation scientists and conservationists are stakeholders too.

3 **Problem definition.** Characterise and delimit the problem for which conservation concepts are to be used.

The perception of the exact nature of the problem at hand may not be same for different conservationists and/or different other stakeholders. Discussion and some sort of agreement is necessary to avoid misunderstandings.

4 **Alternative concepts.** Different potential conservation approaches should be considered.

This step consists in stepping back for a moment from one's own favoured ideas of conservation and human–nature relationships and becoming familiar with other potential concepts, and with the human–nature relationship they imply. This requires a differentiated view on the concepts and narratives, in their existing variety. It must go beyond simple dichotomies and avoid exaggerated depictions of the human–nature relationships implied in the respective concepts. Human–nature relationships, in a narrative form (as developed in Section 3.2), may be a better point of departure than the concepts or specific value types and objects. Actors should be aware of their held ideas of good human–nature relationships, possibly distinct from that foregrounded for mainly strategic reasons. Strategic use of concepts should be handled cautiously and with awareness of potential drawbacks. Multiple, at best combined human–nature relationships should be allowed for. Compatibility of different concepts could be discussed, e.g., in the context of conservation narratives and the tables presented in Section 3.4.

5 **Context.** The biophysical and the social contexts of the situation in question have to be analysed.

The biophysical setting must be analysed as it constrains potential paths of development of ecosystems and social-ecological systems and thus the spectrum of possible actions and outcomes.

On the social side, specifications of how different groups are affecting conservation or are being affected by it are necessary. Analyse power relations as well as potential winners and losers of conservation for the application of each specific concept. Scrutinise also which parts of non-human nature (especially the potential target objects of all stakeholders) are affected by the application of specific conservation concepts. Include as many actants as possible in the process of analysis.

6 **Human–nature relationships.** The differences in what is considered as good human–nature relationships by different stakeholders are to be characterised.

There are different methods to analyse and compare such differences and related conservation goals and potential conflicts. Most popular are currently different methods of *values assessments* (see Section 3.4), where the different kinds of values that groups of people hold with respect to nature are analysed. The rich literature on *discourse analysis* is also relevant here. There is of course the option to explore relationships by extending the scheme for narrative analysis of human–nature relation in Section 3.2 beyond the immediate conservationist community for which it is intended up to now. Stress should be given especially on the *desired* roles of humans with respect to non-human nature. In all cases special consideration should be given to non-material relations (relational approach, as discussed in Section 5.3).

7 **Towards solutions.** Search for win-win solutions in selecting conservation concepts, or otherwise good compromises, but accept that not all conflicts may be solved.

The ideal is that the different ideas of good human–nature relationships present in the specific case will either be harmonised in the same place or by spatial separation within a larger area (as in Biosphere Reserves). Discuss potential solutions under the perspective of justice. Win-win solutions are not always possible. Iterations may be necessary, with other conservation approaches considered, also after scrutinising the biophysical setting or the (desired) human–nature relationships of different stakeholders. Nevertheless, conflicts may not always be solved. Avoid some kinds of overly "integrative" approaches that in fact may hide existing conflicts or become too vague by trying to include everything. There is no unique methodology for this, but a rich literature exists on conflict management in conservation and environmental protection (see previous section).

This is a very simple procedural scheme (much simpler than political scientists might build it…). Yet, given constraints of time and means, it will not always be possible to follow all the steps, especially not with involving stakeholders from the beginning. But even without being able to involve a larger set of stakeholders, the scheme should be used. My main point here is that for selecting the most adequate conservation concepts and for arriving at good human–nature relationships for a broad array of people, it should be paramount for conservationists to be reflexive about the various issues discussed here and to account for the all aspects described.

In the final chapter that follows, I will summarise the results of this book, draw some general conclusions, and provide a very brief outlook.

Notes

1 "Frames" or "framings" are a broader concept than "narrative". Framings constitute the general interpretative background, the cognitive and societal context for describing situations, whether conscious or not. They may or may not be expressed as narratives. See Lejano et al. (2013, p. 53ff) for a discussion of the differences between narratives, frames, and discourses.

2 "Even in the best scenarios of conventional conservation practice the losses [of biodiversity] should be considered unacceptable by civilized people" (Wilson 2016, p. 187).
3 "Handle so, dass du die Menschheit sowohl in deiner Person, als in der Person eines jeden anderen jederzeit zugleich als Zweck, niemals bloß als Mittel brauchst" (Grundlegung der Metaphysik der Sitten, 429).
4 Volume 35.
5 Many thanks to Uta Eser by whom I was inspired to look at Buber's writings.
6 "Die Welt als Erfahrung gehört dem Grundwort Ich-Es zu. Das Grundwort Ich-Du stiftet die Welt der Beziehung" (German original 1923).
7 www.cbd.int/article/cop15-final-text-kunming-montreal-gbf-221222. Accessed May 9, 2023.
8 Exceptions are approaches like that of the "Voluntary Human Extinction Movement", with an emphasis on the *voluntary* (www.vhemt.org/aboutvhemt.htm).
9 As a good example, see, e.g., the framework developed by Chan et al. (2012) for using the ecosystem services concept in management and planning.

References

Adams, W.M. (2004). *Against extinction: the story of conservation*. Earthscan, London, New York.
Agrawal, A. (1997). The politics of development and conservation: legacies of colonialism. *Peace & Change*, 22, 463–482.
Agrawal, A. and Redford, K. (2009). Conservation and displacement: an overview. *Conservation and Society*, 7, 1–10.
Bauer, T. (2018). *Die Vereindeutigung der Welt. Über den Verlust an Mehrdeutigkeit und Vielfalt*. Reclam, Stuttgart.
Berghöfer, U., Rode, J., Jax, K. et al. (2022). 'Societal Relationships with Nature': a framework for understanding nature-related conflicts and multiple values. *People and Nature*, 4, 534–548.
Berghöfer, U., Rozzi, R. and Jax, K. (2010). Many eyes on nature–diverse perspectives in the Cape Horn Biosphere Reserve and their relevance for conservation. *Ecology and Society*, 15(1): 18. [online] URL: www.ecologyandsociety.org/vol15/iss1/art18/
Bridgewater, P. (2016). The man and biosphere programme of UNESCO: rambunctious child of the sixties, but was the promise fulfilled? *Current Opinion in Environmental Sustainability*, 19, 1–6.
Brockington, D. and Igoe, J. (2006). Eviction for conservation: a global overview. *Conservation and Society*, 4, 424–470.
Buber, M. (1970/1973). *I and Thou*. Trans. by Walter Kaufmann. Charles Scribner's Sons, New York. German original 1923.
Büscher, B. and Fletcher, R. (2020). *The conservation revolution: radical ideas for saving nature beyond the Anthropocene*. Verso, London, New York.
Cafaro, P., Butler, T., Crist, E. et al. (2017). If we want a whole Earth, nature needs half. *Oryx*, 51, 400.
Chakrabarty, D. (2009). The climate of history: four theses. *Critical Inquiry*, 35, 197–222.
Chan, K.M.A., Balvanera, P., Benessaiah, K. et al. (2016). Why protect nature? Rethinking values and the environment. *Proceedings of the National Academy of Sciences of the United States of America*, 113, 1462–1465.
Chan, K.M.A., Guerry, A.D., Balvanera, P. et al. (2012). Where are cultural and social in ecosystem services? A framework for constructive engagement. *BioScience*, 62, 744–756.

Chan, K.M.A., Pringle, R.M., Ranganathan, J. et al. (2007). When agendas collide: human welfare and biological conservation. *Conservation Biology*, 21, 59–68.

Chan, K.M.A. and Satterfield, T. (2013). Justice, equity, and biodiversity. In: *Encyclopedia of biodiversity* (ed. Levin, S.). Elsevier, Amsterdam, Vol. 4, pp. 434–442.

Douglas, M. (1966). *Purity and danger: an analysis of concepts of pollution and taboo.* Routledge, London, New York.

Dowie, M. (2009). *Conservation refugees: the hundred-year conflict between global conservation and native people.* MIT Press, Cambridge, Massachusetts.

Dudley, N. (2013). *Guidelines for applying protected area management categories including IUCN WCPA best practice guidance on recognising protected areas and assigning management categories and governance types.* IUCN, Gland.

Duffy, R. (2014). Waging a war to save biodiversity: the rise of militarized conservation. *International Affairs*, 90, 819–834.

Ehrlich, P. and Ehrlich, A. (1981). *Extinction: the causes and consequences of the disappearance of species.* Random House, New York.

Ehrlich, P.R. (1980). The strategy of conservation, 1980–2000. In: *Conservation biology – an evolutionary – ecological perspective* (eds. Soulé, M.E. and Wilcox, B.A.), pp. 329–344.

Escobar, A. (2016). Thinking-feeling with the earth: territorial struggles and the ontological dimension of the epistemologies of the South. *AIBR Revista de Antrropología Iberoamericana*, 11, 11–32.

Eser, U. (2002). Der Wert der Vielfalt: »Biodiversität« zwischen Wissenschaft, Politik und Ethik. In: *Umwelt–Ethik–Recht* (eds. Bobbert, M., Düwell, M. and Jax, K.). Francke-Verlag, Tübingen, pp. 160–181.

Eser, U., Neureuther, A.-K., Seyfang, H. et al. (2014). *Prudence, justice and the good life: a typology of reasoning in selected European national biodiversity strategies.* Bundesamt für Naturschutz, Bonn. https://portals.iucn.org/library/node/44639.

Gordon, G. (2018). Environmental personhood. *Columbia Journal of Environmental Law*, 43, 49–91.

Gould, R.K., Pai, M., Muraca, B. et al. (2019). He 'ike 'ana ia i ka pono (it is a recognizing of the right thing): how one indigenous worldview informs relational values and social values. *Sustainability Science*, 14, 1213–1232.

Haila, Y., Henle, K., Apostolopoulou, E. et al. (2014). Confronting and coping with uncertainty in biodiversity research and praxis. *Nature Conservation*, 8, 45–75.

Himes, A. and Muraca, B. (2018). Relational values: the key to pluralistic valuation of ecosystem services. *Current Opinion in Environmental Sustainability*, 35, 1–7.

Hoelle, J., Gould, R.K. and Tauro, A. (2022). Beyond 'desirable' values: expanding relational values research to reflect the diversity of human–nature relationships. People and Nature. https://doi.org/10.1002/pan3.10316

Hutton, J., Adams, W.M. and Murombedzi, J.C. (2005). Back to the barriers? Changing narratives in biodiversity conservation. *Forum for Development Studies*, 32, 341–370.

Ingold, T. (2000). *The perception of the environment: essays on livelihood, dwelling and skill.* Routledge, London.

Jax, K., Barton, D.N., Chan, K.M.A. et al. (2013). Ecosystem services and ethics. *Ecological Economics*, 93, 260–268.

Jax, K., Calestani, M., Chan, K.M.A. et al. (2018). Caring for nature matters: a relational approach for understanding nature's contributions to human well-being. *Current Opinion in Environmental Sustainability*, 35, 22–29.

Jax, K. and Heink, U. (2015). Searching for the place of biodiversity in the ecosystem services discourse. *Biological Conservation*, 191, 198–205.

Kopnina, H. (2016). Half the earth for people (or more)? Addressing ethical questions in conservation. *Biological Conservation*, 203, 176–185.

Lejano, R., Ingram, M. and Ingram, H. (2013). *The power of narrative in environmental networks*. MIT Press, Cambridge/Massachusetts.

Mace, G.M. (2014). Whose conservation? *Science*, 345, 1558–1560.

Matulis, B.S. and Moyer, J.R. (2017). Beyond Inclusive conservation: the value of pluralism, the need for agonism, and the case for social instrumentalism. *Conservation Letters*, 10, 279–287.

McShane, T.O., Hirsch, P.D., Trung, T.C. et al. (2011). Hard choices: making trade-offs between biodiversity conservation and human well-being. *Biological Conservation*, 144, 966–972.

Meine, C. (2020). From the land to socio-ecological systems: the continuing influence of Aldo Leopold. *Socio-Ecological Practice Research*, 2, 31–38.

Miller, D. (2021). Justice. In: *The Stanford encyclopedia of philosophy (Fall 2021 edition)* (ed. Zalta, E.N.) https://plato.stanford.edu/archives/fall2021/entries/justice/

Miller, M. (2019). Environmental personhood and standing for nature: examining the Colorado River case. *The University of New Hampshire Law Review*, 17, 355–377.

Millstein, R.L. (2018). Debunking myths about Aldo Leopold's land ethic. *Biological Conservation*, 217, 391–396.

Muradian, R., Arsel, M., Pellegrini, L. et al. (2013). Payments for ecosystem services and the fatal attraction of win-win solutions. *Conservation Letters*, 6, 274–279.

Muradian, R. and Gómez-Baggethun, E. (2021). Beyond ecosystem services and nature's contributions: is it time to leave utilitarian environmentalism behind? *Ecological Economics*, 185, 107038.

Norton, B.G. (1991). *Toward unity among environmentalists*. Oxford University Press, New York.

Nussbaum, M.C. (2022). *Justice for animals: our collective responsibility*. Simon & Schuster, New York.

Obrador-Pons, P. (2007). Dwelling. In: *Companion encyclopaedia of geography: from local to global* (eds. Douglas, I., Huggett, R. and Perkins, C.), pp. 1–12. Routledge, London.

Paloniemi, R., Apostolopoulou, E., Cent, J. et al. (2015). Public participation and environmental justice in biodiversity governance in Finland, Greece, Poland and the UK. *Environmental Policy and Governance*, 25, 330–342.

Pascual, U., Balvanera, P., Christie, M. et al. (eds.) (2022). *Summary for policymakers of the methodological assessment report on the diverse values and valuation of nature of the Intergovernmental Science-Policy Platform on Biodiversity and Ecosystem Services*. IPBES secretariat, Bonn.

Paulson, N., Laudati, A.N.N., Doolittle, A. et al. (2012). Indigenous peoples' participation in global conservation: looking beyond headdresses and face paint. *Environmental Values*, 21, 255–276.

Peterson, A.L. (2001). *Being human: ethics, environment and our place in the world*. University of California Press, Berkeley, Los Angeles.

Redpath, S.M., Gutiérrez, R.J., Wood, K.A. et al. (2015a). An introduction to conservation conflicts. In: *Conflicts in conservation: navigating towards solutions* (eds. Redpath, S.M., Gutiérrez, R.J., Wood, K.A. et al.). Cambridge University Press, Cambridge, pp. 3–18.

Redpath, S.M., Gutiérrez, R.J., Wood, K.A. et al. (eds.) (2015b). *Conflicts in conservation: navigating towards solutions*. Cambridge University Press, Cambridge.

Redpath, S.M., Young, J., Evely, A. et al. (2013). Understanding and managing conservation conflicts. *Trends in Ecology & Evolution*, 28, 100–109.

Rogers, A., Castree, N. and Kitchin, R. (eds.) (2013). *Dictionary of human geography (online)*. Oxford University Press, Oxford.

Sandbrook, C. (2017). Weak yet strong: the uneven power relations of conservation. *Oryx*, 51, 379–380.

Sandbrook, C., Fisher, J.A., Holmes, G. et al. (2019). The global conservation movement is diverse but not divided. *Nature Sustainability*, 2, 316–323.

Sangha, K.K., Le Brocque, A., Costanza, R. et al. (2015). Ecosystems and indigenous wellbeing: an integrated framework. *Global Ecology and Conservation*, 4, 197–206.

Sarkar, S. (1999). Wilderness preservation and biodiversity conservation – keeping divergent goals distinct. *BioScience*, 49, 405–412.

Shoreman-Ouimet, E. and Kopnina, H. (2015). Reconciling ecological and social justice to promote biodiversity conservation. *Biological Conservation*, 184, 320–326.

Simberloff, D. (2004). Community ecology: is it time to move on? *American Naturalist*, 163, 787–799.

Strang, V. (2005). Knowing me, knowing you: aboriginal and European concepts of nature as self and other. *Worldviews*, 9, 25–56.

Takacs, D. (1996). *The idea of biodiversity: philosophies of paradise*. John Hopkins University Press, Baltimore, London.

Trepl, L. (1987). *Geschichte der Ökologie. Vom 17. Jahrhundert bis zur Gegenwart*. Athenäum, Frankfurt/Main.

Warren, K.J. (2015). Feminist environmental philosophy. In: *The Stanford encyclopedia of philosophy (Summer 2015 edition)* (ed. Zalta, E.N.) https://plato.stanford.edu/entries/feminism-environmental/

Wiggins, B.J., Ostenson, J.A. and Wendt, D.C. (2012). The relational foundations of conservation psychology. *Ecopsychology*, 4, 209–215.

Wilson, E.O. (ed.) (1988). *Biodiversity*. National Academy Press, Washington D.C.

Wilson, E.O. (2016). *Half Earth: our planet's fight for life*. Liveright Publication Corporation, New York.

Wittmer, H., Rauschmayer, F. and Klauer, B. (2006). How to select instruments for the resolution of environmental conflicts? *Land Use Policy*, 23, 1–9.

6
CONCLUSIONS AND OUTLOOK

The objective of this book was to bring some order into the multitude of conservation concepts. The aim was to support conservation by searching for clearer options on the question: *how do we want to form (or reform) our relationship with nature as humans – and which conservation approaches can help us?*

In Chapter 3, the task of analysing the differences between conservation concepts was approached from different angles. In principle, the concepts may be ordered along the objects they aim at, the values they embrace, or the narratives contained in them. As my starting point I chose to use the latter, the implicit *narratives* of (good) human–nature relationships, because they integrate objects, values, ideas of nature, and even practical aspects of being in and with nature. Based on an analytical scheme that I developed for the analysis of such human narratives, I distinguished 6½ narratives, which I see as major ones in conservation concepts, the half one being "making nature whole again" (restoration), which mostly builds on the other six narratives. In addition, I ordered a larger number of conservation concepts and what I called "supporting concepts" into "conceptual clusters". These clusters assemble concepts that are closely related content-wise. Most clusters are coarsely aligned to one of the narratives distinguished before. A closer look on some of the individual concepts revealed the great breadth and variation of target objects and of value types attributed to concepts within a cluster. Given the variation of meanings of the concepts, often even for one and the same term, it is extremely difficult or up to impossible to unambiguously attribute clearly particular values or particular target objects to most of the major concepts such as ecosystem services, wilderness, or biodiversity in a one-to-one relation. That also means that all to simple dichotomies between conservation concepts are not mirroring the real complexities of the respective concepts. A broader, more integrating path towards selecting appropriate conservation concepts may thus be via analysing the concepts

DOI: 10.4324/9781003251002-6

with respect to the major narratives of human–nature relationships contained, which may implicitly or explicitly guide conservationists' approaches.

An essential background for differentiating between types of human–nature relationships is the concept of nature itself, which has been discussed in Chapter 4. I tried to show that even in Western realm, where current conservation has its origin, multiple ideas of nature and the natural exist at the same time. These ideas, as I have emphasised repeatedly, are intimately related to human self-understanding, i.e. the very conception(s) of what it means to be human, as humans define themselves both in contrast as in relation to nature. Together, ideas of nature and the human define the potential roles of humans in nature, as reflected in the different narratives. While the narratives of human–nature relationships described are those found within the community of conservationists and conservation scientists, others exist within Western culture as well as in non-Western ones, not the least indigenous societies. Given that the global South and also other "less-developed" and seeming "wild" areas are a major target of global conservation efforts (e.g. by establishing protected areas), care has to been taken that Western ideas of nature and human–nature relationships are not taken for granted and/or simply imposed on people living in these areas. This is also a matter of justice, as elaborated in Chapter 5. A common "we" seldom exists, not among people affected by or affecting conservation nor within the community of conservationists and conservation scientists. Questions of justice (and, related to it, of power) hint again at the intimate connection between understandings of nature and human self-understandings. Good relationships between humans and nature need to consider both relations to non-human nature as well as those to other humans. Other humans' well-being is affected by the way individuals deal with nature, and relations with nature affect the personal identity of individual humans as well as that of human communities, and, of course, the well-being of non-human organisms. In what I have called a relational imperative, I postulated that conservation should place major emphasis on what has been called "relational values" or, even broader, relational approaches to human–nature relationships. Relational values mean valuing not just the objects themselves, e.g. for their economic values, but valuing the very – non-material – relations, e.g. of a traditional landscape providing a feeling of home. They thus come close to most peoples' intuition about the values of nature, also those of most indigenous peoples. Going beyond instrumental and intrinsic values, they are not to replace them but are a necessary addition.

Coming back to searching for appropriate conservation concepts in order to arrive at good human–nature relationships, I argued that there is neither one perfect concept nor relationship with nature to be favoured as a unifying approach. Although concepts should not be proliferated without necessity, I am pleading for making use of the existing variety of concepts, adapted to specific settings. Some criteria for selecting appropriate conservation concepts have been formulated in Chapter 5 as well as suggestions of how to put them into practice.

So what are the major conclusions to stress at the end of this book?

First of all, conservation concepts are messy. It is difficult to sort them in a clear and useful way. To some degree this is normal. Concepts develop, are being refined, but often also diluted so that they become fuzzy ("ecosystem services" and "biodiversity" are prominent examples for all of it). Messiness is also a sign of a dynamic and developing research landscape, and not for every purpose are precise definitions necessary or useful. Nevertheless, in instances of putting concepts into practice, and already when they are contested and debated, some rigour of what concepts mean and reflexivity of their use is needed. This is what this book wants to support. The approaches developed earlier, the scheme for narrative analysis, the narratives of human–nature relationships, the conceptual clusters, and the analysis of specific conservation concepts by this means are suggestions. They are open to refinement and further development.

Increased reflexivity in conservation is necessary. There is still too much infighting between different directions of conservations. Not all disagreements can be harmonised, by far not. But I think if conservationists avoid purity, look deeper into the complexity of concepts and refrain from creating and attacking exaggerated depictions of the opponents' positions, much more unity, more compromise, and a bundling of efforts would be possible. Including broader accounts of values, and also the use of narratives as a tool, will help to bring out parts of dissent that lie in deep convictions about worldviews, about the nature of nature, the role of humans within nature, and what it means to be human.

As described in Chapter 3, the last stages of the narratives on human–nature relationships refer to the anticipated future, to visions of how good human–nature relationships might or should look like. I do not have a full-fledged vision nor a detailed pathway towards the future. But there are some elements of such a vision, or wishes, which I want to share.

My first wish of where conservationists should come together is to unite about the need of *respect* towards nature, especially towards all other living beings. This should be the starting point for *all* conservation: respect as an attitude and one that guides how we deal with "the other" – human as well as non-human. We should perceive of these "others" not just as an "it" but also a "thou", to use Buber's terminology. This is also the basis for an attitude of care. I believe that this is in fact what many, if not most, conservationist and conservation scientists already embrace. But it should also be expressed more clearly as an important argument in public, as an important part of good human–nature relationships – linking, in addition, also easily to indigenous perspectives on the world. In a world, where even respect among humans very often appears to be a problem, this sounds somewhat utopian. But that's what we need and should strive for.

Humility is also a thing that comes to my mind here. Humility is not weakness, even though it is often portrayed as such. Humility is not an attitude of subjugation or obedience to authorities. It is an attitude that is aware of one's own limitations, limitations to control nature and other people, of our embeddedness in multiple relations, and thus of humanness. Humility does also not mean to ignore power relations, neither those between individuals nor those transmitted by structures as, e.g., the prevailing economic system and its main beneficiaries. Humility, including the realisation that there is neither one "true" nature nor one idea of good human–nature relationships only, helps to make discourses about the future of nature more rational, without neglecting how much conservation is also about emotions. With regard to nature, humility requires not to negate differences between humans and other species but to negate the idea of a fundamental dichotomy between humans and the rest of nature and in particular a total superiority of humans in and over nature.

If I had a second wish to express, I would emphasise once more that a plurality of conservation concepts should be used. Concepts like those of cultural landscapes, or that of novel ecosystems as potential objects of conservation, have largely been neglected in the mainstream of conservation, or – in the latter case – even been vilified. Protecting nature, protecting species (as the most conspicuous expression of "biodiversity") is not limited to wilderness protection and the promotion of ecosystem services as the two extremes, also not to protected areas. (Re-) discovering nature in novel ecosystems, even as being wild there, is not a substitute for other conservation concepts (and was never meant by its proponents), but it can be a valuable addition in many places – to extend conservation far beyond the classical protected areas, even in urban spaces. Likewise, cultural landscapes, if they are accepted as an object of conservation at all, are not necessarily static and to be maintained in a museum-like fashion. Also cultural landscapes develop; as with other conservation concepts, it is a matter of societal negotiation (and the specific ecological conditions) what kind of development is possible, is acceptable, and what is not, in social terms as from a conservation perspective. So, I personally love the wild but I do not subscribe to a pure "Humans apart from nature" relationship. I love cultural landscapes and the associated idea of "Being at home in nature" but I do want to turn it into a museum. Compromises, good compromises, are possible.

Looking into the future, we should proceed from a crisis experience to positive visions of the future. Both positive and negative, sometimes even apocalyptical, visions for the future exist, often as alternative futures of a narrative, and both types of vision can be made productive for further research and for thinking about conservation. It is, most of the time, easier to develop negative visions than specific positive ones. But positive visions are direly needed; they are formulated in repeatedly updated goals in institutions like the United Nations, the CBD, IPBES, the European Union, and many national states. But at best they should be developed also on a regional and local scale, jointly by a diverse set of conservationists together with the local people, who search for common visions of how they want to live from, in, and with nature.

INDEX

Note: Figures are indicated by *italics*. Tables are indicated by **bold**. Endnotes are indicated by the page number followed by 'n' and the endnote number e.g., 20n1 refers to endnote 1 on page 20.

actants 69, 71–2
Agrawal, Arun 15, 225, 230, 232
Amazonia 178–82
animal protection 11, 35, 64–6, 89
Anangu people 184
Anderson, Jay E. 157
Angermeier, Paul L. 157, 159
Anthropocene concept 82–3, 129
anthropocentrism 46, 59, 60, 106, 220
anthropology 38, 155, 161, 174
archaeobiota 159; *see also* exotic species
Ascension island 97
Australia: aboriginal cultures 184–6, 228; country as narrative 184–6 (*see also* songlines); and exotic species 186
Aymara people 200

Berghöfer, Uta 161
Berchtesgadener Land 240
Berkes, Fikret 175, 177, 178, 182–3, 203–4
balance of nature 43, 74, 163
biocultural conservation 118–19, **136**
biodiversity 89, 107–10, **134**, 216–18; functional diversity 108–9, **134**; typical diversity 109–10
bird protection history of 11–13
Biosphere Reserves 218, 240–1, 248
Bolivia 197–200

boundary object 2–3, 216, 217
Brundtland, Gro Harlem 35–6
Buber, Martin 237
buen vivir 176, 198–201
Büscher, Bram 160, 229–31

Callicott, J. Baird 171–2, 181, 183, 188
Campa people 179, 181
care 60, 88–9, 96, 183, 186, 186, 193, 255
Carson, Rachel 30, 93–4
Chakrabarty, Dipesh 222–3
Christian Religion *see* Judeo-Christian religion
colonialism 14–16, 87, 223–6, 232
concepts 39–40; descriptive and normative 39; dynamics of 102–7; and power 39–40; and terms 39; thick concepts 41
conceptual clusters; cluster "biodiversity" 107–10, **134**; cluster "cultural landscapes" 114–19, **136**; cluster "ecosystem functioning" (supporting concepts) 129–31; cluster "ecosystem management" 119–22, **137**; cluster "ecosystem services" 103–7, **133**; cluster "novel ecosystems" 122–5, **138**; cluster "restoration" 125–8, **139**; cluster "wilderness" 110–14, **135**; compared 131–40; defined 102–3

conservation; and capitalism 229–32; and colonialism (*see* colonialism); and constructivism (*see* constructivism); of cultural landscapes (*see* cultural landscapes); civilisation argument 10, 14, 25, 87, 225, 230; definition 8; and economy 12, 47, 83, 109, 229–31; and human well-being (*see* human well-being); heritage argument 14, 17, 19, 25, 89, 115, 117, 225; of individual organisms 11, 64–6; international institutions and agreements 13–14, 27 (*see also* IPBES, CBD, IUCN, World Heritage Convention); NGOs 27, 223, 226; and patriotism or nationalism 17–19; related fields 34–7; scientification of 43–4, 216; social context 40–2, 222–33; strategical dimensions 57, 62, 243–4; visions for 81, 255–6

conservation biology; as a crisis discipline 77; founding 31; characterisation 37, 44–9, *45*; as a normative science 40–1

conservation conflicts 220, 221, 233, 238–40, 246, 248

conservation concepts; compared 131–40; definition 39; origins and dynamics 57; pluralism 219–22, 244–6, 256; proliferation of 218–19; purposes of use 234, 246; strategic use 105–9, 243–4; succession of 1–4, 214–16; supporting concepts 120, 129–31

conservation, history of 4, 9–32; 19th and early 20th century 9–26; after 1945 26–32

conservation narratives (major) 85–102; accounting for multiple 244–6; "being at home in nature" 92–4; "caring for other living beings ..." 88–91; compared 99–102, *101*; "gardening nature in novel ways" 95–7; "humans apart from nature" 86–8; "making nature whole again" 97–9; "safeguarding useful and necessary nature" 91–2; "steering nature in a responsible way" 94–5

conservation objects 57–9; pragmatic and ultimate 58

conservation science 44–8, *45*

constructivism 165–74

Convention on Biological Diversity (CBD) 27, 31, 108, 241

convivial conservation 230

Conwentz, Hugo 17–18

cosmology 177–9, 184–5, 191

Costanza, Robert 104
countryside, access to 23
creation *see* cosmology
crisis 77–8, 80, 83, 84, 86, **100**, 191, 203–4
Costanza, Robert 104–5
Cree people 182–4
Cronon, William 165–6, 168, 170–1, 174
cultural ecology 37–8
cultural landscapes 22–5, 93, 114–19, *116*, 126, 132, **136**, 156, 181, 188, 190, 240, 256

Daily, Gretchen 67–8, 92
designer ecosystems 123, **138**
Diaz, Sandra 196–7
Drachenfels mountain 16–17, *16*
dreamings and dreaming tracks *see* songlines
dwelling perspective 161, 236

ecocentrism 59
ecological integrity 120, **137**
ecological values 61
ecosystem, definitions 130
ecosystem approach 2, 119–20
ecosystem-based management *see* ecosystem management
ecosystem functioning 65, 130–1, 242
ecosystem health 120, **137**
ecosystem management 94, 119–21, **137**
ecosystem services 31, 91–2, 103–7, 133; *see also* nature's contributions to people (NCPs); cultural services 133; and biodiversity 105; and ethics 226–9; history of concept 103–6
Ecuador 200
Ehrlich, Paul and Anne 67, 103–4, 232
environmental protection 30, 35
environmental ethics 30, 32, 59–61, 65, 174
Eifel Highlands 126
Eser, Uta 61
eudaimonia 60–1
exotic species 109, 158–9, 186
extinction of species 13, 31, 67, 70, 82, 90, 216

fallacy of disembedded ideas 192
feminist ethics 161, 200, 236
Fletcher, Robert 160, 229–31
flux of nature 43, 74, 163
fortress conservation 223, 225, 241
framing 214–15, *215*, 248n1
furusato 190

gardening metaphor 97, 128
gifts of nature 199–200, 205n9
good human life 60–1, 89, 236; *see also* eudaimonia

Hacking, Ian 168
Half Earth 113–14, 230
Hawai'i 204
healing metaphor 98, 128
Heimat 23–4, **136**, 190
historical fidelity 127; *see also* restoration
Hobbs, Richard 122
Homeland 23, **136**
human ecology 37
human embeddedness in nature 161, 193, 236, 255
human-nature dichotomy *see* human-nature dualism
human-nature dualism 160, 162, 173, 191, 199–200
human-nature relationships: *see also* "narratives" and "conservation narratives"; accounting for multiple 244–8; non-western 174–94, 199–204, 236; potential fabula of its history 79–81; problems of "translation" between cultures 192, 194, 199–204; reciprocity 178, 181–3
humans: (presumed) superiority over nature 160, 164, 181, 256; self-understanding as related to nature 34, 154, 164, 173, 213, 223, 234–5, 238, 254
human well-being 45–8, 91–2, *104*, 105, 201, 227–8
humility 183, 255–6
hybrid ecosystems 123–4, **138**

ideal types 84, 141n6
identity, personal and social 23–5, 93, 117, 164, 183–5, 236, 242, 254
imperialism *see* colonialism
indigenous and local knowledge (ILK) *see* traditional ecological knowledge
indigenous people 14–16, 72, 174–8, 199, 202, 225, 255; myths about environmental behaviour 175–6
Ingold, Tim 161, 236
Intergovernmental Platform on Biodiversity and Ecosystem Services (IPBES) 31, 194–204; conceptual framework 194–202, *195*; and ILK 202; "translation" between scientific and other knowledge 196, 199, 200, 202–4

International Union for the Conservation of Nature (IUCN), history 27

Japan 187–90
Judeo-Christian religion 191–2
justice 61, 223–9, 238–40; environmental and ecological 61, 224, 239; procedural 224

Kant, Immanuel 236
Kareiva, Peter 44–8
Kirchhoff, Thomas 115–16
knowledge forms 119, 178, 194–202, 244

landscape concept 114–15
Latour, Bruno 168
Lease, Gary, 165–6, 172
Leopold, Aldo 28, *29*, 68, 94, 98, 121–2, 125, 245–6
Leopold, Aldo Starker ("Leopold Report") 127
Levy-Strauss (structure of myths) 141n4
local ecological knowledge 175; *see also* traditional ecological knowledge

Mace, Georgina 214–15
Makuna people 180–1
Malawi principles 119–20
Man and the Biosphere Programme 27, 119, 218
Marsh, George Perkins 18, 25, 214
Marvier, Michelle 44–8
McCord, Edward 90–1
metaphors 140n3
Mother Earth (Pachamama) 195–201, *195*
Millennium Ecosystem Assessment 104–5, *104*, 201
Muir, John 21, 25, *26*

Nachhaltigkeit see sustainability
naive realism 167
narrative analysis 66–8; *see also* conservation narratives
narratives: definition of 68–70; fabula and story 68–9, 81, 84; main elements 69; structure of narratives on human-nature relationships ("stages") 70–1, 73–9; as tool to "translate" between ideas of human-nature relationships 202–4
national parks 19–22; eviction of native people (*see also* colonialism) 21, 87, 226
Native Americans 21, 87; *see also* Cree people

nativeness 109, 157–9, 186
Naturdenkmal see natural monument
natural/naturalness: *see also* nature; as benchmark for conservation 65–6, 155–9; descriptive and normative 155–8
natural capital 105, **133**; *see also* ecosystem services
natural heritage 14–19, 22–6, 89, 141n8, 115–16, **134**, **136**
natural monument 16–18, 89, 107, **134**
natural resource management 18–19, 35, 91–2
naturalistic fallacy 157–8
nature: *see also* natural: concept of 33–4, 65, 154–8, 160–5, 174, 197; as construct (*see* constructivism); dichotomies 154–5, 160–2; edenic ideas about 14, 25; pristine 87; rights of 193, 198, 200, 208, 224, 239; and science 33–4; underuse of 183
nature-based solutions 105, **133** (*see also* ecosystem services)
nature's contribution to people (NCPs) 106, **133**, 217, 219; *see also* ecosystem services
Naturschutz 23–4
Navarino island (Chile) 221–2
"new conservation" 44–9
neobiota *see* exotic species
no-analog ecosystems *see* novel ecosystems
novel ecosystems 96, 122–5, **138**, 218, 256; critique of 124–5
Nussbaum, Martha 201, 236

Oostvaardersplaasen 62–6, *64*

Pachamama *see* "Mother Earth"
personhood of non-human nature 180, 189, 193
physiocentrism *see* ecocentrism
Pinchot, Gifford 19, 36, 214
place; and space 242; sense of 93, 117, 183–4, 242–3; significance for conservation 222
political ecology 38
power relations 41–2, 176, 223–9, 231, 233, 247, 254
preservation 36
progress, idea of 74–6, 225–6
Propp, Vladimir ("morphology of the folk tale") 141n4
protected area categories (IUCN) 18, 22, 241

protected areas 13–15, 18, 19, 21, 22, 26, 223, 226, 240, 241
Prozessschutz 113
prudential values 61
purity 220, 245, 250, 255

reciprocity *see* human-nature relationships
reclamation **139**; *see also* restoration
rehabilitation **139**; *see also* restoration
relational approaches 60–1, 234–8, 248, 254; *see also* values, relational
relativism, environmental 171–3
religion and nature 177–8, 187, 191–2
reflexivity in conservation 255
remediation **139**; *see also* restoration
resilience 120
respect for nature 178, 181–3, 193, 255
restoration 98, 125–8, **139**, 242; relation to conservation 36; roles of history 127–8
rewilding 63–5, 112, **135**; and animal protection 63–5
Ritter, Carl 115
Roosevelt, Theodore 114, 117
Royal Society for the Prevention of Cruelty to Animals (RSPCA) 11
Rudorff, Ernst 23–5, *26*

Sami people 180, 182, 183
Sarasin, Paul 27
satoyama 187–90
St. Clair River 95
Schweitzer, Albert 193
science and conservation 26, 28–32, 42–4, 201–2, 216–18, 240, 242–3
semi-natural area **136**
social-ecological systems 94, 120–1, **137**
societal relationships with nature (SRN) 161
Society for Restoration Ecology (SER) 125; guildelines for restoration 126, 128
socio-ecological systems 120–1, **137**
songlines 185–6
Soulé, Michael 31, 37, 44–8, 77, 165–8, 170
Spaemann, Robert 155
species protection 89–91
stakeholder involvement 119, 126, 246–8
stewardship 96, 129, 192
sustainability; origins of idea 19; management for sustainability 35–6

Takacs, David 217
Taylor, Paul 193

Thoreau, Henry David 88
totems 185
traditional ecological knowledge (TEK) 119, 174–8; *see also* local ecological knowledge; terminology 175
Tukano people 181

value, definition of 59; values of nature 57, 59–62; instrumental 59–61; intrinsic 59–61; relational 60–61, 234–7, 254
Vera, Frans 63, 65–6
vivir bien see buen vivir
"Voices of the poor" study 105, 201

Weber, Max 141n6
"Western" and "non-Western" (defined) 41–2

White Lynn Jr. 191
wildlife protection; colonial origins 13–16
wilderness 22, 87, 110–14, 135, 153, 167, 181, 220–2, 241, 245; critique of 110, 113–14; history of concept 110; Indian wilderness 87; Wilderness Act (USA) 22, 110–12
wildness 65–6, 110, 112–14, **135**
Wilson, Edward O. 109, 113, 229, 249n2
World Heritage Convention 18, 115–16
Wulaia (Chile) 221–2

Yaghan people 221
Yellowstone National Park 19, *20*, 87, 119

Zahneiser, Howard 111

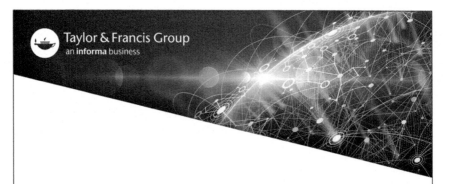